The Handbook of Environmental Chemistry

Volume 5 Water Pollution
Part F, Volume 3

O. Hutzinger
Editor-in-Chief

Environmental Impact Assessment of Recycled Wastes on Surface and Ground Waters

Volume 3
Engineering Modeling and Sustainability

Volume Editor: Tarek A. Kassim

With contributions by
J. Atmadja · A. C. Bagtzoglou
C. V. Chrysikopoulos · W. D. Constant
P. P. Feng · M. N. Goltz · M. R. Ilic · T. A. Kassim
R. R. Kommalapati · J. W. Park · P. S. Polic †
K. T. Valsaraj · C. J. Werth · H. C. Young

Springer

Volume Editor

Professor Dr. T. A. Kassim
Department of Civil and
Environmental Engineering
Seattle University
901 12th Avenue
Seattle, WA 98122-1090
USA
kassimt@seattle.edu

Library of Congress Control Number: 2004112898

ISSN 1433-6863
ISBN-10 3-540-23585-x Springer-Verlag Berlin Heidelberg New York
ISBN-13 978-3-540-23585-x
DOI 10.1007/b101937

Springer is a part of Springer Science+Business Media
springeronline.com

Printed in Germany

Cover Design: E. Kirchner, Springer-Verlag
Production Editor: Christiane Messerschmidt, Rheinau
Typesetting: Fotosatz-Service Köhler GmbH, Würzburg
Printed on acid-free paper 52/3141 – 5 4 3 2 1 0

Editor-in-Chief

Advisory Board

The Handbook of Environmental Chemistry Also Available Electronically

For all customers who have a standing order to The Handbook of Environmental Chemistry, we offer the electronic version via SpringerLink free of charge. Please contact your librarian who can receive a password for free access to the full articles by registering at:

springerlink.com

If you do not have a subscription, you can still view the tables of contents of the volumes and the abstract of each article by going to the SpringerLink Homepage, clicking on "Browse by Online Libraries", then "Chemical Sciences", and finally choose The Handbook of Environmental Chemistry.

You will find information about the

- Editorial Bord
- Aims and Scope
- Instructions for Authors
- Sample Contribution

at springeronline.com using the search function.

Preface

Environmental Chemistry is a relatively young science. Interest in this subject, however, is growing very rapidly and, although no agreement has been reached as yet about the exact content and limits of this interdisciplinary discipline, there appears to be increasing interest in seeing environmental topics which are based on chemistry embodied in this subject. One of the first objectives of Environmental Chemistry must be the study of the environment and of natural chemical processes which occur in the environment. A major purpose of this series on Environmental Chemistry, therefore, is to present a reasonably uniform view of various aspects of the chemistry of the environment and chemical reactions occurring in the environment.

The industrial activities of man have given a new dimension to Environmental Chemistry. We have now synthesized and described over five million chemical compounds and chemical industry produces about hundred and fifty million tons of synthetic chemicals annually. We ship billions of tons of oil per year and through mining operations and other geophysical modifications, large quantities of inorganic and organic materials are released from their natural deposits. Cities and metropolitan areas of up to 15 million inhabitants produce large quantities of waste in relatively small and confined areas. Much of the chemical products and waste products of modern society are released into the environment either during production, storage, transport, use or ultimate disposal. These released materials participate in natural cycles and reactions and frequently lead to interference and disturbance of natural systems.

Environmental Chemistry is concerned with reactions in the environment. It is about distribution and equilibria between environmental compartments. It is about reactions, pathways, thermodynamics and kinetics. An important purpose of this Handbook, is to aid understanding of the basic distribution and chemical reaction processes which occur in the environment.

Laws regulating toxic substances in various countries are designed to assess and control risk of chemicals to man and his environment. Science can contribute in two areas to this assessment; firstly in the area of toxicology and secondly in the area of chemical exposure. The available concentration ("environmental exposure concentration") depends on the fate of chemical compounds in the environment and thus their distribution and reaction behaviour in the environment. One very important contribution of Environmental Chemistry to the above mentioned toxic substances laws is to develop laboratory test methods, or mathematical correlations and models that predict the environmental fate of new chemical compounds. The third purpose of this Handbook is to help in the basic understanding and development of such test methods and models.

The last explicit purpose of the Handbook is to present, in concise form, the most important properties relating to environmental chemistry and hazard assessment for the most important series of chemical compounds.

At the moment three volumes of the Handbook are planned. Volume 1 deals with the natural environment and the biogeochemical cycles therein, including some background information such as energetics and ecology. Volume 2 is concerned with reactions and processes in the environment and deals with physical factors such as transport and adsorption, and chemical, photochemical and biochemical reactions in the environment, as well as some aspects of pharmacokinetics and metabolism within organisms. Volume 3 deals with anthropogenic compounds, their chemical backgrounds, production methods and information about their use, their environmental behaviour, analytical methodology and some important aspects of their toxic effects. The material for volume 1, 2 and 3 was each more than could easily be fitted into a single volume, and for this reason, as well as for the purpose of rapid publication of available manuscripts, all three volumes were divided in the parts A and B. Part A of all three volumes is now being published and the second part of each of these volumes should appear about six months thereafter. Publisher and editor hope to keep materials of the volumes one to three up to date and to extend coverage in the subject areas by publishing further parts in the future. Plans also exist for volumes dealing with different subject matter such as analysis, chemical technology and toxicology, and readers are encouraged to offer suggestions and advice as to future editions of "The Handbook of Environmental Chemistry".

Most chapters in the Handbook are written to a fairly advanced level and should be of interest to the graduate student and practising scientist. I also hope that the subject matter treated will be of interest to people outside chemistry and to scientists in industry as well as government and regulatory bodies. It would be very satisfying for me to see the books used as a basis for developing graduate courses in Environmental Chemistry.

Due to the breadth of the subject matter, it was not easy to edit this Handbook. Specialists had to be found in quite different areas of science who were willing to contribute a chapter within the prescribed schedule. It is with great satisfaction that I thank all 52 authors from 8 countries for their understanding and for devoting their time to this effort. Special thanks are due to Dr. F. Boschke of Springer for his advice and discussions throughout all stages of preparation of the Handbook. Mrs. A. Heinrich of Springer has significantly contributed to the technical development of the book through her conscientious and efficient work. Finally I like to thank my family, students and colleagues for being so patient with me during several critical phases of preparation for the Handbook, and to some colleagues and the secretaries for technical help.

I consider it a privilege to see my chosen subject grow. My interest in Environmental Chemistry dates back to my early college days in Vienna. I received significant impulses during my postdoctoral period at the University of California and my interest slowly developed during my time with the National Research Council of Canada, before I could devote my full time of Environmental Chemistry, here in Amsterdam. I hope this Handbook may help deepen the interest of other scientists in this subject.

Amsterdam, May 1980 *O. Hutzinger*

Twentyone years have now passed since the appearance of the first volumes of the Handbook. Although the basic concept has remained the same changes and adjustments were necessary.

Some years ago publishers and editors agreed to expand the Handbook by two new open-end volume series: Air Pollution and Water Pollution. These broad topics could not be fitted easily into the headings of the first three volumes. All five volume series are integrated through the choice of topics and by a system of cross referencing.

The outline of the Handbook is thus as follows:

1. The Natural Environment and the Biochemical Cycles,
2. Reaction and Processes,
3. Anthropogenic Compounds,
4. Air Pollution,
5. Water Pollution.

Rapid developments in Environmental Chemistry and the increasing breadth of the subject matter covered made it necessary to establish volume-editors. Each subject is now supervised by specialists in their respective fields.

A recent development is the accessibility of all new volumes of the Handbook from 1990 onwards, available via the Springer Homepage springeronline.com or springerlink.com.

During the last 5 to 10 years there was a growing tendency to include subject matters of societal relevance into a broad view of Environmental Chemistry. Topics include LCA (Life Cycle Analysis), Environmental Management, Sustainable Development and others. Whilst these topics are of great importance for the development and acceptance of Environmental Chemistry Publishers and Editors have decided to keep the Handbook essentially a source of information on "hard sciences".

With books in press and in preparation we have now well over 40 volumes available. Authors, volume-editors and editor-in-chief are rewarded by the broad acceptance of the "Handbook" in the scientific community.

Bayreuth, July 2001 *Otto Hutzinger*

Contents

Contents of Volume 1 XIII

Contents of Volume 2 XIV

Foreword XV

Equilibrium Partitioning and Mass Transfer of Organic Chemicals Leached from Recycled Hazardous Waste Materials
C. J. Werth 1

Organic Chemicals in Groundwater: Modeling Fate and Transport
M. N. Goltz · J.-W. Park · P. P. Feng · H. C. Young 33

Mathematical Methods for Hydrologic Inversion: The Case of Pollution Source Identification
A. C. Bagtzoglou · J. Atmadja 65

Nonaqueous Phase Liquid Pool Dissolution in Subsurface Formations
C. V. Chrysikopoulos 97

A Case Study in the Application of Environmental Chemodynamic Principles for the Selection of a Remediation Scheme at a Louisiana Superfund Site
R. R. Kommalapati · W. D. Constant · K. T. Valsaraj 133

Solidification/Stabilization Technologies for the Prevention of Surface and Ground Water Pollution from Hazardous Wastes
M. R. Ilic · P. S. Polic † 159

Waste Minimization and Molecular Nanotechnology: Toward Total Environmental Sustainability
T. A. Kassim 191

Subject Index 231

Contents of Volume 1

Environmental Impact Assessment of Recycled Wastes On Surfaces and Ground Waters Concepts; Methodology and Chemical Analysis

Volume Editors: Tarek A. Kassim and Kenneth J. Williamson
ISBN 3-540-00268-5

Environmental Impact Assessment: Principles, Methodology and Conceptual Framework
T. A. Kassim · B. R. T. Simoneit

Recycling Solid Wastes as Road Construction Materials: An Environmentally Sustainable Approach
T. A. Kassim · B. R. T. Simoneit · K. J. Williamson

Beneficial Reuses of Scrap Tires in Hydraulic Engineering
R. R. Gu

Hazardous Organic Chemicals in Biosolids Recycled as Soil Amendments
A. Bhandari · K. Xia

A Review of Roadway Water Movement for Beneficial Use of Recycled Materials
D. S. Apul · K. H. Gardner · T. T. Eighmy

Evaluation Methodology for Environmental Impact Assessment of Industrial Wastes Used as Highway Materials: An Overview with Respect to U.S. EPA's Environmental Risk Assessment Framework
P. O. Nelson · P. Thayumanavan · M. F. Azizian · K. J. Williamson

Leaching from Residues Used in Road Constructions – A System Analysis
D. Bendz · P. Flyhammar · J. Hartlén · M. Elert

Forensic Investigation of Leachates from Recycled Solid Wastes: An Environmental Analysis Approach
T. A. Kassim · B. R. T. Simoneit · K. J. Williamson

Contents of Volume 2

Environmental Impact Assessment of Recycled Wastes
On Surfaces and Ground Waters
Risk Analysis

Volume Editor: Tarek A. Kassim
ISBN 3-540-23587-6

Using Laboratory Experiments and Computer Models for Assessing the Potential Risk of Recycled Waste Materials – Case Studies
D. Rasmussen · M. Winther-Nielsen · D. Graham · B. Halling-Sørensen

Environmental Impacts of Leachate from Portland Cement Concrete (PCC) with and Without Plasticizer in Highway Construction
M. F. Azizian · P. O. Nelson · P. Thayumanavan · K. J. Williamson

Environmental Impact Assessment of Lignite Fly Ash and Its Utilization Products as Recycled Hazardous Wastes on Surface and Ground Water Quality
P. S. Polic † · M. R. Ilic · A. R. Popovic

Application of Whole Effluent Toxicity Test Procedures for Ecotoxicological Assessment of Industrial Wastes Used as Highway Construction Materials
P. Thayumanavan · P. O. Nelson

Toxicity Evaluation and Human Health Risk Assessment of Surface and Ground Water Contaminated by Recycled Hazardous Waste Materials
R. Rodriguez-Proteau · R. L. Grant

Chemodynamics and Multicontaminant Joint Toxicity Modeling of Organic Leachates from Recycled Solid Wastes
T. A. Kassim · B. R. T. Simoneit

Foreword

Industrial chemicals are essential to support modern society. Growth in the number and quantity of chemicals during recent decades has been extraordinary resulting in an increase in quantity and complexity of hazardous waste materials (HWMs). Many of these HWMs will remain in the environment for long periods of time, which has created a need for new methods for environmentally safe and efficient disposal including recycling and/or reuse of these complex materials. In many areas, existing landfills are reaching capacity, and new regulations have made the establishment of new landfills difficult. Disposal cost continues to increase, while the waste types accepted at solid waste landfills are becoming more and more restricted.

One answer to these problems lies in the ability of industrialized society to develop beneficial uses for these wastes as by-products. The reuse of waste by-products *in lieu* of virgin materials can relieve some of the burdens associated with disposal and may provide inexpensive and environmentally sustainable products. Current research has identified several promising uses for these materials. However, research projects concerning Environmental Impact Assessment (EIA) of various organic and inorganic contaminates in recycled complex mixtures and their leachates on surface and ground waters are still needed to insure that adverse environmental impacts do not result.

Answers to some of these concerns can be found in the present book, entitled "Environmental Impact Assessment of Recycled Wastes on Surface and Ground Waters". This book is an attempt to comprehensively understand the potential impacts associated with recycled wastes. The book is divided into three main volumes, each with specific goals.

The first volume of the book is subtitled "Concepts, Methodology and Chemical Analysis". It focuses on impact assessment and decision-making in project development and execution by presenting the general principles, methodology and conceptual framework of any EIA investigation. It discusses various sustainable engineering applications of industrial wastes, such as the reuses of various solid wastes as highway construction and repair materials, scrap tires in hydraulic engineering projects, and biosolids as soil amendments. It also evaluates several chemical and ecotoxicological methodologies of waste leachates, and introduces a unique "forensic analysis and genetic source partitioning" modeling technique, which consists of an environmental "molecular marker" approach integrated with various statistical/mathematical modeling tools. In addition, several case studies are presented and discussed, which: (a) provide comprehensive information of the interaction between hydrology and solid wastes incorporated into highway materials; (b) assess potential ecological risks posed by constituents released from waste and industrial by-products used in highway construction; and (c) describe the processes and events that are crucial for assessing the contaminant leaching from roads where residues are used as construction material by using interaction matrices.

The second volume of the book, subtitled "Risk Analysis", is problem-oriented and includes several multi-disciplinary case studies. It evaluates various experimental methods and models for assessing the risks of recycling waste products, and ultimately presents the applicability of two hydrological models such as MIKE SHE and MACRO. This volume is also background information-oriented, and presents the principles of ecotoxicological and human risk assessments by: (a) discussing the use of the whole effluent toxicity (WET) tests as predictive tools for assessing ecotoxicological impacts of solid wastes and industrial by-products for use as highway materials; (b) providing information on the concepts used in estimating toxicity and human risk and hazard due to exposure to surface and ground waters contaminated from the recycling of hazardous waste materials; and (c) introducing an advanced modeling approach that combines the physical and chemical properties of contaminants, quantitative structure-activity and structure-property relationships, and the multicomponent joint toxic effect in order to predict the sorption/desorption coefficients, and contaminant bioavailability.

The third volume of the book is subtitled "Engineering Modeling and Sustainability". It presents, examines and reviews: (a) the fundamentals of important chemodynamic (i.e., fate and transport) behavior of environmental chemicals and their various modeling techniques; (b) the equilibrium partitioning and mass transfer relationships that control the transport of hazardous organic contaminants between and within highway construction materials and different phases in the environment; (c) several physical, chemical, and biological processes that affect organic chemical fate and transport in ground water; (d) simulation models of organic chemical concentrations in a contaminated ground water system that vary over space and time; (e) mathematical methods that have been developed during the past 15 years to perform hydrologic inversion and specifically to identify the contaminant source location and time-release history; (f) various case studies that demonstrate the utility of fate and transport modeling to understand the behavior of organic contaminants in ground water; (g) recent developments on non-aqueous phase liquids (NAPL) pool dissolution in water saturated subsurface formations; and (h) correlations to describe the rate of interface mass transfer from single component NAPL pools in saturated subsurface formations.

In addition, this volume examines various hazardous waste treatment/disposal and minimization/prevention techniques as promising alternatives for sustainable development, by: (a) presenting solidification/stabilization treatment processes to immobilize hazardous constituents in wastes by changing these constituents into immobile (insoluble) forms; binding them in an immobile matrix; and/or binding them in a matrix which minimizes the material surface exposed to weathering and leaching; (b) providing an overview of waste minimization and its relationship to environmental sustainability; (c) portraying the causes of sustainability problems and diagnosing the defects of current industrial manufacturing processes in light of molecular nanotechnology; and (d) analyzing and extrapolating the prospect of additional capabilities that may be gained from the development of nanotechnology for environmental sustainability.

It is important to mention that information about EIA of recycled wastes on surface and ground waters is too large, diverse, and multi-disciplined, and its knowledge base is expanding too rapidly to be covered in a single book. Nevertheless, the authors tried to present the most important and valid key principles that underlie the science and engineering aspects of risk analysis, characterization, and assessment.

It is hoped that the present information help the reader continue to search for creative and economical ways to limit the release of contaminants into the environment, to develop highly sensitive techniques to track contaminant once released, to find effective methods to remediate contaminated resources, and to promote current efforts toward promoting environmental sustainability.

Seattle, Washington, USA
March, 2005

Tarek A. Kassim

Handb Environ Chem Vol. 5, Part F, Vol. 3 (2005): 1–32
DOI 10.1007/b11493

Equilibrium Partitioning and Mass Transfer of Organic Chemicals Leached from Recycled Hazardous Waste Materials

Charles J. Werth (✉)

Department of Civil and Environmental Engineering, University of Illinois at Urbana-Champaign, 205 N. Mathews Ave., Urbana, IL 61801, USA
werth@uiuc.edu

1	**Introduction**	2
2	**Equilibrium Partitioning**	3
2.1	Overview	3
2.2	Thermodynamics of Equilibrium Partitioning	4
2.3	Air-Water Equilibrium Relationship	5
2.4	Octanol-Water Equilibrium Relationship	9
2.5	Solid-Water Equilibrium Relationship	10
2.6	Comparison of Fugacity Capacities	12
3	**Mass Transfer**	13
3.1	Overview	13
3.2	Thermodynamics of Diffusion	14
3.3	Diffusion Coefficients	15
3.3.1	Diffusion in Water	15
3.3.2	Diffusion in Vapor	16
3.3.3	Diffusion in Porous Media	17
3.4	Transient Diffusion	18
3.4.1	Fick's Second Law	18
3.4.2	Solutions of Fick's Second Law	20
3.4.3	Diffusion in Natural Solids	23
3.5	Mass Transfer Between Phases	24
3.5.1	Film Theory	24
3.5.2	Boundary Layer Theory	26
3.5.3	Empirical Correlations	28
4	**Conclusions**	29
	References	30

Abstract Potentially hazardous waste materials (HWMs) are increasingly being recycled and used as highway construction and repair materials (CRMs). While reducing disposal costs, this practice raises concerns because hazardous organic pollutants (HOPs) from these wastes can leach from highways and enter soil surface and ground waters. This chapter presents the equilibrium partitioning and mass transfer relationships that control the transport of HOPs between and within highway CRM and different phases in the environment. Partitioning relationships are derived from thermodynamic principles for air, liquid, and solid phases, and they are used to determine the driving force for mass transfer. Mass transfer relationships are developed for both transport within phases, and transport between phases. Some analytical solutions for mass transfer are examined and applied to relevant problems.

Keywords Partitioning · Soil · Water · Sorption · Diffusion · Interphase mass transfer · Thermodynamics

List of Abbreviations

CRM	Construction and repair material
FSG	Fuller-Schettler-Giddings
HOP	Hazardous organic pollutant
HWM	Hazardous waste material
L	Length
M	Mass
NAPL	Nonaqueous phase liquid
PAH	Polycyclic aromatic hydrocarbon
PCB	Polycyclic chlorinated biphenyl
RAP	Reclaimed asphalt pavement
t	Time
T	Temperature
TCE	Trichloroethene
VOC	Volatile organic chemical

1 Introduction

Various types of recycled and potentially hazardous waste materials (HMWs) are used as highway construction and repair materials (CRMs). Used tires are shredded and used as fill material for parking lots and roads. Reclaimed asphalt pavement (RAP) is broken up and incorporated into new asphalt, and used as fill material. Both of these recycled materials are petroleum based. Hence, they are made from a suite of organic compounds including polycyclic aromatic hydrocarbons (PAHs) and volatile organic compounds (VOCs). Toxicologists have shown that at least some PAHs and VOCs are potentially carcinogenic [1]. Since these hazardous organic pollutants (HOPs) can leach from roadways [2], their fate in the environment is of primary concern.

This chapter focuses on the equilibrium partitioning and mass transfer of HOPs in the environment. HOPs of concern can be classified as either semi-volatile or volatile, and this distinction is important when determining the fate of chemicals in the environment. The environment consists of various media including the HWM, ground water, surface water, air, soil, sediment, and vegetation. From the perspective of equilibrium partitioning and mass transfer relationships, these different media can often be categorized as vapor, liquid, or solid. In the second section of this chapter thermodynamic principles are used to derive equilibrium partitioning relationships between different phases. In the third section a mass balance approach is used to define mass transfer relationships both within and between phases, and various mass transfer rate constant expressions are presented. The conclusions of this chapter are presented in the fourth section. This chapter also contains several examples that illustrate how equilibrium partitioning and mass transfer relationships can be used to evaluate the fate of HOPs in the environment.

2 Equilibrium Partitioning

2.1 Overview

Figure 1 illustrates the different phases available for organic chemicals in the environment. These may include the original HWM, as well as water, air, soil, and other organic chemicals present in liquid or solid phases. As previously mentioned, the HWM may be recycled-asphalt pavement or shredded tires. Both the HWM and soil can be very complex. HWM may consist of different polymers, rocks and mineral fragments, and various HOPs. Soils consist of different types and amounts of natural organic matter, black carbon, and minerals. As discussed below, carbonaceous materials typically have the greatest affinity for HOPs.

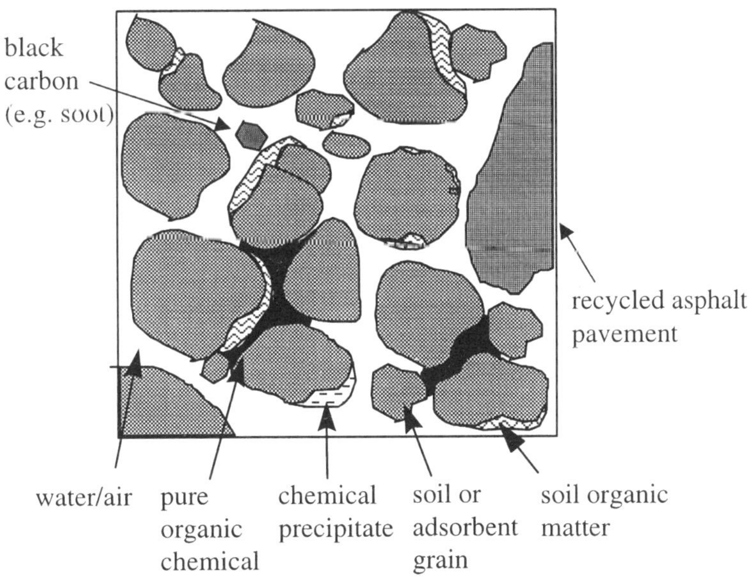

Fig. 1 Different phases available for partitioning in the environment

2.2
Thermodynamics of Equilibrium Partitioning

The Gibbs free energy, G [Energy M^{-1}], is often used to develop relationship that can be used to determine whether two phases are in equilibrium. The change in the Gibbs free energy, ΔG, denotes whether a phase transfer is favorable or not. For example, when a HOP is transferred from phase A to B the ΔG for this phase change is negative. The change in the Gibbs free energy with respect to the amount of mass transferred is defined as the chemical potential:

$$\mu_i = \left(\frac{\partial G}{\partial n_i}\right)_{T,P,n_j} \quad (1)$$

where n_i is the moles of i in the system at constant temperature T, pressure P, and moles of j (n_j). When the chemical potential in one phase is equal to the chemical potential in another phase, the two phases are in equilibrium. More explicitly, in a system consisting of π phases and m hazardous waste species, the conditions for equilibrium are

$$T^{(1)} = T^{(2)} = \dots T^{(\pi)} \quad (2)$$

$$P^{(1)} = P^{(2)} = \dots P^{(\pi)} \quad (3)$$

$$\mu_1{}^{(1)} = \mu_1{}^{(2)} = \dots \mu_1{}^{(\pi)} \quad (4)$$

.

$$\mu_m{}^{(1)} = \mu_m{}^{(2)} = \dots \mu_m{}^{(\pi)} \quad (5)$$

For an ideal gas, the change in the chemical potential for an isothermal change from pressure P_i^o to P_i at constant temperature T is

$$\mu_i - \mu_i^o = RT \ln (P_i/P_i^o) \quad (6)$$

where R is the ideal gas constant [e.g., 0.08206 atm l $Kelvin^{-1}$ mol^{-1}], P_i^o is the pressure of compound i at a reference state, and μ_i^o is the chemical potential of i at this same reference state. To generalize to real gases, liquids, and solids, Lewis [3] defined a function f, called fugacity, such that for an isothermal change

$$\mu_i - \mu_i^o = RT \ln (f_i/f_i^o) \quad (7)$$

where f_i^o is the fugacity of i at the reference state. For an ideal gas $f_i=y_iP$, where y_i is the mol fraction of i and P is the total pressure.

Now consider two phases α and β:

$$\mu_i^{\alpha} - \mu_i^{o,\alpha} = RT \ln (f_i^{\alpha}/f_i^{o,\alpha}) \quad (8)$$

$$\mu_i^{\beta} - \mu_i^{o,\beta} = RT \ln (f_i^{\beta}/f_i^{o,\beta}) \quad (9)$$

If the reference state is chosen the same in α and β then

$$\mu_i^{o,\alpha} = \mu_i^{o,\beta} \quad (10)$$

At equilibrium the chemical potentials are equal:

$$\mu_i^\alpha = \mu_i^\beta \tag{11}$$

It then follows without any loss of generality that

$$f_i^\alpha = f_i^\beta \tag{12}$$

So at equilibrium, the fugacity of any species must be equal in phases α and β. This result is general, in that the fugacity of any species must be equal in all phases at equilibrium. For example, if benzene (B) is distributed between soil, water, and air at equilibrium, then the fugacity of benzene in each of these phases is identical.

In order to determine the equilibrium partitioning relationship between two phases, the fugacity relationship for each of the two phases must be set equal. Fugacity is expressed in units of pressure. It can be thought of as an escaping tendency. If the fugacity for a HWM in phase α is greater than phase β, then the HWM escapes from phase α to β until equilibrium is achieved (i.e., the escaping tendencies in the two phases are equal). To determine the fugacity for phases other than air, it is convenient to define the fugacity in terms of the concentration in a particular phase:

$$f_i = C_i/Z_i \tag{13}$$

where C_i is the concentration of the chemical of interest [M L^{-3}] in any phase and Z_i is the fugacity capacity of i in the same phase [M L^{-3} pressure^{-1}]. Mackay and Paterson [4] derived values of Z_i for each phase in the environment, and they used these expressions to derive equilibrium partitioning relationships. In a similar way, the equilibrium relationships between different phases are derived from the corresponding fugacity relationships in the next section.

2.3 Air-Water Equilibrium Relationship

HOPs from HWMs can find their way into surface water and ground water. The persistence of these chemicals in water can depend on how easily they can volatilize from solution. To derive the air-water equilibrium relationship we start with the fugacity relationship for the air phase:

$$f_{i,\,air} = \phi_i x_i P = \phi_i P_i = C_{i,\,air}/Z_{i,\,air} \tag{14}$$

where ϕ_i is the fugacity coefficient [-] and x_i is the mole fraction of species i [-]. The parameter ϕ_i accounts for the nonideality of the air phase. For organic chemicals of concern in the environment, ideal behavior can often be assumed. Hence, ϕ_i=1 and the ideal gas law can be used to define P_i as follows:

$$P_i = n_i RT/V_i \tag{15}$$

where V_i represents the volume [L^3] of i. Substituting Eq. (15) into Eq. (14), the fugacity capacity can be defined:

$$Z_{i,air} = C_{i,air} V_i/(n_i RT) = 1/RT \quad (16)$$

With Eq. (16) the fugacity can be defined in terms of the concentration in air:

$$f_{i,air} = x_i P_T = P_i = C_{i,air} RT \quad (17)$$

The next step is to define the fugacity in water. The initial expression for water is similar to that for air:

$$f_{i,water} = \gamma_i x_i P_i^{sat} = C_{i,water}/Z_{i,water} \quad (18)$$

where γ_i is the liquid-phase activity coefficient [-] and P_i^{sat} is the saturation vapor pressure of pure i at temperature T. When x_i approaches 1, then γ_i=1 and f_i is equal to the product of x_i and P_i^{sat}. This is analogous to Rauolt's law. For hydrophobic organic chemicals in water, x_i is typically much less than 1 (infinitely dilute), and the relationship between x_i and γ_i is generally of the form [4]

$$\ln \gamma_i = K(1 - x_i)^2 \quad (19)$$

where K=constant. Since x_i is small, ln γ_i~K. Hence, γ_i is relatively constant. This assumption yields

$$f_{i,water} = \gamma_i x_i P_i^{sat} = K' x_i P_i^{sat} = C_{i,water}/Z_{i,water} \quad (20)$$

where K'=exp(K)=constant. Equation (20) can be rearranged to define $Z_{i,water}$:

$$Z_{i,water} = C_{i,water}/(K' x_i P_i^{sat}) \quad (21)$$

$$v_{m,water} = x_i/C_{i,water} = \text{molar volume of water (L/mol)}$$

$$Z_{i,water} = 1/(K' v_{m,water} P_i^{sat}) = I/H_i$$

where H_i is the Henry's constant [e.g., atm l mol^{-1}]. This allows us to define the fugacity as follows:

$$f_{i,water} = C_{i,water} H_i \quad (22)$$

Setting the fugacities for air (Eq. 17) and water (Eq. 22) equal to each other, the equilibrium relationship for these two phases is derived:

$$C_{i,air} RT = C_{i,water} H_i \quad (23)$$

$$H_{cc,i} = H_i/RT = C_{i,water}/C_{i,air} \quad (24)$$

where $H_{cc,i}$ is the dimensionless Henry's constant [-]. Alternatively, we could have used the fugacity expression for air defined in terms of the partial pressure to obtain the dimensional form of the Henry's constant equation as follows:

$$P_i = C_{i,water} H_i \quad (25)$$

$$H_i = P_i/C_{i,water} \quad (26)$$

Both Eqs. (24) and (26) are different versions of Henry's Law. Henry's Law is valid for predicting air-water equilibria for many organic pollutants of concern. Henry's law is valid in the following range:

$$P \sim 1 \text{ atm.}; T = 10 \rightarrow 60\,°C, x_i < 0.001$$

Table 1 Henry's constant values at 20 °C

Compound	H_i (atm m^3 gmol^{-1})	Compound	H_i (atm m^3 gmol^{-1})
Nonane	0.332	Tetrachloroethylene	0.0141
n-Hexane	0.883	Trichloroethylene	0.00842
2-Methylpentane	0.633	Tetralin	0.00136
Cyclohexane	0.140	Decalin	0.106
Chlorobenzene	0.00341	Vinyl chloride	0.0217
1,2-Dichlorobenzene	0.00168	Chloroethane	0.0110
1,3-Dichlorobenzene	0.00294	Hexachloroethane	0.00591
1,4-Dichlorobenzene	0.00259	Carbon tetrachloride	0.0232
o-Xylene	0.00474	1,3,5-Trimethylbenzene	0.00571
p-Xylene	0.00645	Ethylene dibromide	0.000610
m-Xylene	0.00598	1,1-Dichloroethylene	0.0218
Propylbenzene	0.00881	Methylene chloride	0.00244
Ethylbenzene	0.00601	Chloroform	0.00332
Toluene	0.00555	1,1,2,2-Tetrachloroethane	0.000730
Benzene	0.00452	1,2-Dichloropropane	0.00190
Methyl ethylbenzene	0.00503	Dibromochloromethane	0.00103
1,1-Dichloroethane	0.00563	1,2,4-Trichlorobenzene	0.00183
1,2-Dichloroethane	0.00147	2,4-Dimethylphenol	0.0101
1,1,1-Trichloroethane	0.0146	1,1,2-Trichlorotrifluoroethane	0.245
1,1,2-Trichloroethane	0.000740	Methyl ethyl ketone	0.000190
cis-1,2-Dichloroethylene	0.00360	Methyl isobutyl ketone	0.000290
trans-1,2,-Dichloroethylene	0.00857	Methyl cellosolve	0.116

Reproduced from [5].

Table 1 lists several compounds and their Henry's constants taken from Ashworth et al. [5]. For compounds of similar structure, heavier compounds tend to have smaller Henry's constant values. For compounds of similar size, those with polar functional groups (e.g., oxygen, nitrogen, sulfur) tend to have smaller Henry's constant values. This explains why methyl ethyl ketone (MW=72) has a Henry's constant that is orders of magnitude less than chloroethane (MW=64).

When Henry's constants are not available, it is often adequate to calculate the Henry's constant from the saturation vapor pressure of the pure liquid and the aqueous phase solubility limit, $C_{i,sat}$, as follows:

$$H_i = P_{i,sat}/C_{i,sat} \tag{27}$$

Several authors have investigated the temperature dependence of Henry's constant for environmentally significant pollutants. Relationships from Ashworth et al. [5] are shown in Table 2.

Numerous methods have also been developed to determine H_i based on molecular connectivity indices (MCIs) and/or polarity descriptors [6].

Table 2 Temperature dependence of Henry's constant values

	H=exp(A–B/T)		
	A	B	r^2
Nonane	–0.1847	202.1	0.013
n-Hexane	25.25	7530	0.917
2-Methylpentane	2.959	957.2	0.497
Cyclohexane	9.141	3238	0.982
Chlorobenzene	3.469	2689	0.965
1,2-Dichlorobenzene	–1.518	1422	0.464
1,3-Dichlorobenzene	2.882	2564	0.850
1,4-Dichlorobenzene	3.373	2720	0.941
o-Xylene	5.541	3220	0.966
p-Xylene	6.931	3520	0.989
m-Xylene	6.280	3337	0.998
Propylbenzene	7.835	3681	0.997
Ethylbenzene	11.92	4994	0.999
Toluene	5.133	3024	0.982
Benzene	5.534	3194	0.968
Methyl ethylbenzene	5.557	3179	0.968
1,1-Dichloroethane	5.484	3137	0.993
1,2-Dichloroethane	–1.371	1522	0.878
1,1,1-Trichloroethane	7.351	3399	0.998
1,1,2-Trichloroethane	9.320	4843	0.968
cis-1,2-Dichloroethylene	5.164	3143	0.974
trans-1,2-Dichloroethylene	5.333	2964	0.985
Tetrachloroethylene	10.65	4368	0.987
Trichloroethylene	7.845	3702	0.998
Tetralin	11.83	5392	0.996
Decalin	11.85	4125	0.919
Vinyl chloride	6.138	2931	0.970
Chloroethane	4.265	2580	0.984
Hexachloroethane	3.744	2550	0.768
Carbon tetrachloride	9.739	3951	0.997
1,3,5-Trimethylbenzene	7.241	3628	0.962
Ethylene dibromide	5.703	3876	0.928
1,1-Dichloroethylene	6.123	2907	0.974
Methylene chloride	8.483	4268	0.988
Chloroform	11.41	5030	0.997
1,1,2,2-Tetrachloroethane	1.726	2810	0.194
1,2-Dichloropropane	9.843	4708	0.820
Dibromochloromethane	14.62	6373	0.914
1,2,4-Trichlorobenzene	7.361	4028	0.819
2,4-Dimethylphenol	–16.34	–3307	0.555
1,1,2-Trichlorotrifluoroethane	9.649	3243	0.932
Methyl ethyl ketone	–26.32	–5214	0.797
Methyl isobutyl ketone	–7.157	160.6	0.002
Methyl cellosolve	–6.050	–873.8	0.023

*Valid from 10 to 30 °C. H, atm m^3 $gmol^{-1}$. T, K.
Reproduced from [5].

2.4 Octanol-Water Equilibrium Relationship

Octanol is a partitioning medium just as water is a partitioning media. While there is nothing inherently special about octanol with respect to other organic liquids, the extent that an organic chemical partitions to octanol from water has become a standard for evaluating hydrophobicity (i.e., chemicals that partition more to octanol from water are more hydrophobic). Since HOPs that are more hydrophobic accumulate more in body tissues, partition more strongly to soils and sediments, and are typically more easily removed by adsorption from water, the extent that HOPs partition to octanol from water is a very important environmental indicator.

The initial fugacity relationship for octanol is similar to that for water:

$$f_{i,oct} = \gamma_{i,oct} x_i P_i^{sat} = C_{i,oct}/Z_{i,oct} \tag{28}$$

where $\gamma_{i,oct}$ is the octanol activity coefficient [-] and the other parameters were defined above. The parameter $\gamma_{i,oct}$ is relatively constant so Eq. (28) can be approximated as follows:

$$f_{i,oct} = Kx_i pP_i^{sat} = C_{i,oct}/Z_{i,oct} \tag{29}$$

where K is a constant. Rearranging, Eq. (29) can be solved for the fugacity capacity:

$$Z_{i,oct} = C_{i,oct}/(Kx_i P_i^{sat}) \tag{30}$$

Since $v_{m,oct}=x_i/C_{i,oct}$ (molar volume of octanol), the parameters that define $Z_{i,oct}$ are constant. Combining constants results in

$$Z_{i,oct} = 1/(K_i v_{m,oct} P_i^{sat}) = 1/K_{i,oct} \tag{31}$$

where $K_{i,oct}$ is the octanol constant. Substituting Eq. (31) into Eq. (28) the fugacity relationship is obtained:

$$f_{i,oct} = C_{i,oct} K_{i,oct} \tag{32}$$

The last step is to set the fugacity expressions for octanol and water equal to each other to obtain

$$C_{i,oct} K_{i,oct} = C_{i,water} H_i \tag{33}$$

Rearranging we obtain the octanol-water partition coefficient [-][1]:

$$K_{i,ow} = H_i/K_{oct} = C_{i,oct}/C_{i,water} \tag{34}$$

Table 3 lists values of K_{ow} for several different compounds. For compounds of similar structure (e.g., hydrocarbons), heavier compounds will generally have greater K_{ow}s. For compounds of similar size, compounds with oxygen or other polar functional groups will have smaller K_{ow}s. For example, the K_{ow} for DDT is

[1] Equation (34) allows us to redefine the fugacity capacity as $Z_{i,ow}=K_{i,ow}/H_i$.

Table 3 Octanol-water partition coefficients[a]

Compound	Log K_{ow}	K_{ow}	Classification
Water	–1.38	0.0417	Hydrophilic
Methanol	–0.77	0.17	Hydrophilic
Propanol	0.3	2	Mildly hydrophobic
Chloromethane	0.91	8.1	Mildly hydrophobic
Chloroform	1.95	89.1	Hydrophobic
TCE	2.29	195	Hydrophobic
Dichlorobenzenes	3.3	1900	Strongly hydrophobic
DDT and PCBs	>5	>100,000	Strongly hydrophobic

[a] CRC [18].

more than 100,000 times greater than the K_{ow} for methanol (methanol is lighter and it has a polar OH group). Hence, it is not surprising that DDT accumulates to a much greater extent in fatty tissue than methanol or other less hydrophobic compounds.

2.5 Solid-Water Equilibrium Relationship

In this section solids represent a partitioning or adsorption phase such as soil, asphalt pavement, or granular activated carbon. In contrast to air, water, and octanol, solid phases are typically very complex and poorly characterized. For example, many studies have shown that soils and sediments are characterized by many different types and amounts of organic matter and minerals, and that these different environments have various affinities for an organic chemical.

The fugacity for the solid or sorbed phase is expressed as follows:

$$f_{i,\,solid} = C_{i,\,solid}/Z_{i,\,solid} = q_{i,\,solid}/Z^{*}_{i,solid} \qquad (35)$$

where $q_{i,solid}$ is the sorbed concentration of i [M M^{-1}], $Z^{*}_{i,solid}$ is a units modified fugacity capacity [M M^{-1} $pressure^{-1}$], and the other parameters were defined previously. Due to the complexity of solids, the relationship between $f_{i,solid}$ and $q_{i,solid}$ (or $C_{i,solid}$) is often not linear, and cannot be obtained directly. Consequently, solid-water equilibrium relationships are obtained independent of the fugacity relationships, and the equilibrium relationships are used to define the fugacity capacity and the fugacity[2]. While any solid-water relationship would do, we initially choose the Freundlich isotherm because of its widespread use with heterogeneous sorbents:

$$q_{i,\,solid} = K_F C_{i,\,water}{}^{n_F} \qquad (36)$$

[2] Instead of a solid-water equilibrium relationship, a solid-air equilibrium relationship could be used in the same manner to define the fugacity capacity and the fugacity.

where K_F is the Freundlich capacity parameter with units of $[(M\,M^{-1})(M\,L^{-3})_F^{-n})]$ and n_F is the dimensionless Freundlich exponent. Setting the fugacity for solid and water equal to each other we obtain

$$q_{i,solid}/Z^*_{i,solid} = C_{i,water} \cdot H_i \tag{37}$$

$$Z^*_{i,solid} = q_{i,solid}/(H_i \cdot C_{i,water}) \tag{38}$$

$$Z^*_{i,solid} = K_F C_{i,water}{}^{n_F - 1}/H_i \tag{39}$$

$$f_{i,solid} = H_i \cdot q_{i,solid}/(K_F C_{i,water}{}^{n_F - 1}) \tag{40}$$

The parameter K_F is a measure of capacity of a solid for a HOP. The parameter n_F is a measure of the favorability of a chemical for a sorbent. When $n_F<1$, sorption at lower concentrations is more favorable than sorption at higher concentrations (i.e., $q_{i,solid}/C_{i,water}$ at low concentrations is greater than $q_{i,solid}/C_{i,water}$ at high concentrations). The opposite is true when $n_F>1$.

For cases when $n_F=1$, the Freundlich equation reduces to a linear isotherm:

$$q_{i,solid} = K_d C_{i,water} \tag{41}$$

where K_d is the equilibrium distribution coefficient $[L^3\,M^{-1}]$. Sorption is often linear over a narrow concentration range. Early sorption studies developed correlations between K_d, K_{ow}, and the fraction of organic carbon, f_{oc} $[M\,M^{-1}]$, such as that in Eq. (42) [7]. Later these correlations were modified to include more information about the sorbent chemical properties, such as that in Eq. (43) [8]:

$$\log(K_d/f_{oc}) = \log(K_{oc}) = 1.00 \log K_{ow} - 0.21 \tag{42}$$

$$\log(K_d/f_{oc}) = \log(K_{oc}) = 1.00 \log K_{ow} - 0.21 + (\log[H]/[O] - 0.73) \tag{43}$$

where K_{oc} is the organic carbon normalized equilibrium distribution coefficient $(L^3\,M^{-1})$ and [H]/[O] represents the atomic ratio of hydrogen to oxygen that comprises the soil organic matter of a sediment.

Many other equilibrium relationships have been applied to model sorption. For example, the Langmuir (Eq. 44) and the Polanyi-Dubinin (Eq. 45) isotherms have been widely applied to adsorption in zeolites [9]:

$$q_{i,solid} = b_i C_{i,water} Q_{i,solid}/(1 + b_i C_{i,water}) \tag{44}$$

$$q_{i,solid} = w_{i,0}\,\rho_i \exp(-a_i(\varepsilon_{i,lw}/V_{i,m})^{di}) \tag{45a}$$

$$\varepsilon_{i,lw} = RT \ln(C_{i,sat}/C_{i,water}) \tag{45b}$$

where i represents the solute, $Q_{i,solid}$ $[M\,M^{-1}]$ and $w_{i,0}$ $[L^3\,M^{-1}]$ represent the mass and volumetric adsorption capacity of the sorbent, respectively, a_i, b_i, and d_i are fitting parameters, ρ_i is the solute density $[M\,L^{-3}]$, $\varepsilon_{i,lw}$ is the solute adsorption potential in water [energy M^{-1}], $V_{i,m}$ is the solute molar volume $[L^3\,mol^{-1}]$, and the remaining parameters were previously defined. While the Freundlich isotherm is empirically based, the Langmuir and Polanyi-Dubinin isotherms are mechanistically based. Use of the Langmuir isotherm assumes that there are a limited number of sorption sites with identical energy and that there are no interactions between neighboring sorbate molecules. Use of the Polanyi-Dubinin isotherm assumes that adsorption occurs via a pore-filling mechanism, where a pure phase of adsorbate forms in the pores and the adsorption potential of pores diminish during filling.

More recently [10, 11], the Langmuir and Polanyi-Dubinin isotherms have been combined with a linear isotherm to describe uptake on soils and sediments. The rationale is that there is both an adsorption component and a partitioning component in natural solids. For example, several authors [12, 13] have presented evidence that supports the hypothesis that organic matter, which typically controls uptake on soils and sediments, is comprised of a rigid or glassy component that controls adsorption and a more flexible or rubbery component that controls partitioning. Other authors [14–16] have presented evidence that suggests a high surface area carbonaceous fraction of the rigid organic matter (often black carbon) controls adsorption. Li and Werth [17] further postulated that it is the micropore spaces in rigid organic matter that control adsorption. See the cited work for a comprehensive treatment of this subject.

2.6 Comparison of Fugacity Capacities

The advantage of using fugacity to calculate the equilibrium distribution coefficients becomes apparent when one compares the fugacity capacities of a HOP for several different phases. For example, consider a region of the unsaturated zone just below the ground surface where naphthalene is distributed between air, water, pure phase octanol, and soil at equilibrium. The fugacity capacities for these phases are repeated below in Eqs. (46)–(49):

$$Z_{i,\mathrm{air}} = 1/RT \tag{46}$$

$$Z_{i,\mathrm{water}} = 1/H_i \tag{47}$$

$$Z_{i,\mathrm{oct}} = K_{i,\mathrm{ow}}/H_i \tag{48}$$

$$Z^{*}_{i,\mathrm{solid}} = K_d/H_i \tag{49}$$

Naphthalene at 25 °C has K_{ow}=2239 and H_i=4.9×10^{-1} atm·l/mol [18]. Chiou et al. [19] measured K_d=6.21 l/kg for naphthalene on Anoka soil. This yields Z_{air}=4.09×10^{-2} mol/(l·atm), Z_{water}=2.04 mol/(l·atm), Z_{oct}=4,569 mol/(l·atm), and $Z^{*}_{i,solid}$=12.7 mol/(kg·atm). It is obvious from these values that most of the naphthalene will reside in the octanol at equilibrium. If octanol is not present, most of the naphthalene will reside on the soil. However, note that the units of the fugacity capacity for soil are different than those for the other phases. Hence, if octanol is not present we can only say that more naphthalene will reside on one kg of soil than one liter of the other phases. An alternative approach would be to convert the units of $Z^{*}_{i,solid}$ by multiplying by the soil bulk density (this allows us to normalize the fugacity capacity by a REV containing soil). If we assume a bulk density equal to 1.75 kg/l [20], $Z_{i,solid}$=22.2 mol/(l·atm). Hence, on a per liter basis most of the mass partitions to the soil in the absence of pure phase octanol.

3 Mass Transfer

3.1 Overview

Figure 2 illustrates several different mass transfer processes in the environment. These include diffusive mass transfer within a phase such as water, air, or solid, and mass transfer between phases.

Mass transfer occurs whenever there is a fugacity gradient. If there is a fugacity gradient within a phase, mass (or a HOP) moves within the phase. If there is a fugacity gradient between phases, mass moves between phases. The stronger the fugacity gradient, the faster mass transfers.

When considering mass transfer within a phase, the fugacity gradient can often be replaced by the concentration gradient without any significant error. When considering mass transfer between phases, the equilibrium relationships derived above can be used to describe the driving force for mass transfer. This will be further explained below. However, first the formal derivation of the driving force for mass transfer is presented.

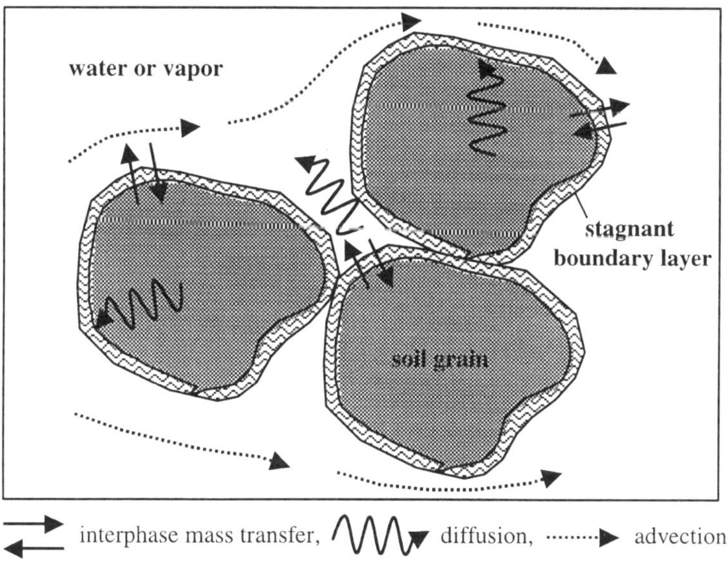

Fig. 2 Illustration of mass transfer processes in the environment

3.2 Thermodynamics of Diffusion

We saw above that at equilibrium both the chemical potential and fugacity must be uniform. It follows that mass transfer occurs when there is a gradient in either chemical potential or fugacity. Consider an element of length dx at constant P, T, and n_j, across which there is a gradient of chemical potential for some species i. Following Atkins [*21*], the maximum work performed by transferring mass across the element is given by

$$dw_i = d\mu_i = \left(\frac{\partial \mu_i}{\partial x}\right)_{P,T,n_j} dx \tag{50}$$

Work can be defined in terms of an opposing force multiplied by distance over which the force is applied. It necessarily follows that Eq. (50) can be redefined in terms of force [M L t^{-2}] as shown below:

$$dw_i = -F_i dx \tag{51}$$

$$F_i = -\left(\frac{\partial \mu_i}{\partial x}\right)_{P,T,n_j} \tag{52}$$

where F_i represents some equivalent thermodynamic force required to move mass across the element. Recall the expression for the chemical potential:

$$\mu_i = \mu_i^o + RT \ln (f_i/f_i^o) \tag{53}$$

Substituting Eq. (53) into Eq. (52) yields

$$F_i = -RT \left(\frac{\partial \ln f_i}{\partial x}\right)_{P,T,n_j} \tag{54}$$

$$F_i = -\frac{RT}{f_i}\left(\frac{\partial f_i}{\partial x}\right)_{P,T,n_j} \tag{55}$$

The fugacity was defined above in terms of a concentration and a fugacity capacity. At constant pressure and temperature, fugacity capacities were constant (except when isotherms were nonlinear). Hence, Eq. (55) can be simplified to the following:

$$F_i = -\frac{RT}{C_i}\left(\frac{\partial C_i}{\partial x}\right)_{P,T,n_j} \tag{56}$$

Diffusing particles experience a viscous drag that opposes F_i. When F_i and the viscous drag are balanced, diffusing particles reach a steady drift speed, s_i (i.e., a steady rate of mass transfer). Hence, F_i is proportional to s_i. Now consider M_i particles that pass through area, A, normal to the direction of transport during time increment Δt. The flux, J_i, of these particles is defined as

$$J_i = \frac{M_i}{A\Delta t} \tag{57}$$

If the concentration of particles in solution, C_i, is constant during Δt, then the flux $[M/(L^2 \cdot t)]$ can also be defined as

$$J_i = C_i s_i \tag{58}$$

Hence, the flux is also proportional to s_i. It follows that J_i must be proportional to the concentration times F_i and thus

$$J_i \propto \frac{\partial C_i}{\partial x} \tag{59}$$

The constant that defines this proportionality is knows as the diffusion coefficient, D_i. Modifying Eq. (59) to include D_i yields Fick's first law:

$$J_i = -D_i \frac{\partial C_i}{\partial x} \tag{60}$$

If transport only occurs in the x direction, Eq. (60) reduces to a normal derivative:

$$J_i = -D_i \frac{dC_i}{dx} \tag{61}$$

3.3 Diffusion Coefficients

Diffusion is the process whereby matter is transported from one part of a system to another as a result of random molecular motion [22]. The random molecular motion is driven by the kinetic energy of individual molecules (i.e. thermal motion). From Eq. (61), the net rate of molecular diffusion is governed by the molecular diffusion coefficient and the concentration gradient. The molecular diffusion coefficient ($D_{i,mol}$) is a property of the solute-solvent system. For very simple systems $D_{i,mol}$ can be derived from first principles. For example, the value of $D_{i,mol}$ for an inert gas can be estimated from kinetic gas theory [21]. For more complex systems, empirical equations are often used.

3.3.1 Diffusion in Water

Several relationships are widely used to calculate values of $D_{i,mol}$ for organic compounds in water and air. One of the most commonly used empirical correlations for predicting the water-phase diffusivity is the Wilke-Chang [23] equation:

$$D_{i,mol} = \frac{7.4 \cdot 10^{-8} (\chi M)^{0.5} T}{\mu V_{i,b}^{0.6}} \tag{62}$$

Table 4 Schroeder Increments for calculating the molar volume

	Increment [cm^3mol^{-1}]		Increment [cm^3mol^{-1}]
Carbon	7	Bromine	31.5
Hydrogen	7	Iodine	38.5
Oxygen	7	Sulfur	21
Nitrogen	7	Double bond	7
Fluorine	10.5	Triple bond	14
Chlorine	24.5	Aromatic ring	–7

Source: Values are from [24].

where $D_{i,mol}$ is the molecular diffusion coefficient (for Eq. 62 the units are [$cm^2\ s^{-1}$]), χ is the association parameter for the solvent (χ=2.6 for water and 1 for nonpolar solvents), $V_{i,b}$ is the molar volume of solute i at its normal boiling pt. (units must be [cm^3/mol]), μ is the liquid viscosity (units must be [centipoise]) (μ=1 centipoise for water), T is the absolute temperature [Kelvin], and M is the molecular wt of the solvent. For simple, non-dissociating organic molecules, the molar volume $V_{i,b}$ can be estimated using Schroeder increments [24]. Table 4 presents common Schroeder increments.

Schroeder increments are additive for an organic compound. For example, V_b for benzene is calculated by summing the contributions from six carbons, six hydrogens, three double bonds, and one aromatic ring. Other methods to calculate the molar volume are discussed in Reinhard and Drefahl [6]. Typical values of $D_{i,mol}$ for organic chemicals in water are on the order of $10^{-5}\ cm^2\ s^{-1}$.

3.3.2
Diffusion in Vapor

In gas, the FSG (Fuller-Schettler-Giddings) method is commonly used to calculate the molecular diffusion coefficient [25]:

$$D_{12,mol} = \frac{10^{-3}T^{1.75}[(M_1 + M_2)/(M_1M_2)]^{0.5}}{P\left(V_{b_1}^{0.33} + V_{b_2}^{0.33}\right)^2} \tag{63}$$

where 1 represents the solvent (air), 2 represents the solute, $D_{12,mol}$ is the molecular diffusivity (again the units must be [cm^2/s]), T is the absolute temperature [Kelvin], P is pressure [atm], M is the molecular weight, V_b is the molar volume at normal boiling point [$cm^3\ mol^{-1}$]. As before, V_b can be calculated from Schroeder Increments.

Typical values of $D_{12,mol}$ for volatile organic chemicals in air are on the order of $10^{-1}\ cm^2$/s. Hence, diffusion coefficients for organics in air are approximately

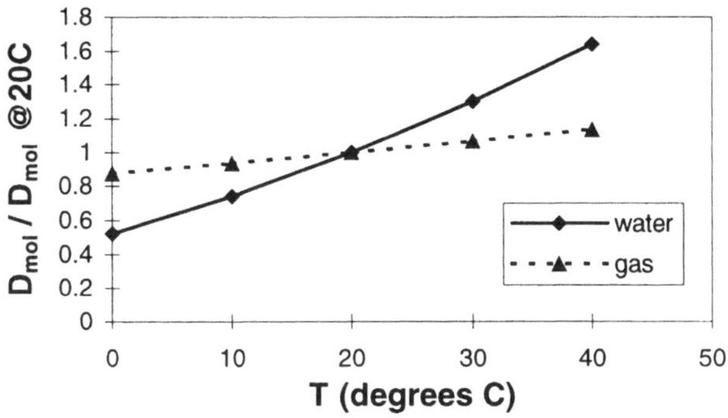

Fig. 3 Temperature dependence of the diffusion coefficient

10,000 times greater than diffusion coefficients in water. An approximate time for diffusion to occur over a distance L can be estimated from

$$t = L^2/D_{mol} \tag{64}$$

Hence, for a fixed distance, diffusion in air occurs about 10,000 times faster than diffusion in water.

The Wilke-Change and FSG equations were used to calculate the temperature dependence of D_{mol} in water and air, respectively. The results (presented as the ratio of D_{mol} to $D_{mol,20°C}$) are shown in Fig. 3. At first the results are counter intuitive because the gas phase appears to be more heavily dependent on T. However, the viscosity of water is a strong function of temperature. Hence, D_{mol} in water is a stronger function of temperature.

3.3.3
Diffusion in Porous Media

The diffusion path is often altered by the presence of solid boundaries. For example, in the subsurface organic chemicals must diffuse around soil and sediment grains. Within soil and sediment grains, organic chemicals must diffuse inside narrow and possibly undulating pores. To account for these effects the effective diffusion coefficient is modified by a restrictivity factor, K_r [-], and a tortuosity factor, τ [-], as follows:

$$D_{eff} = D_{mmol}K_r/\tau \tag{65}$$

The parameter K_r accounts for the effects of steric hindrance on diffusion in small pores. K_r is a strong function of the adsorbate to pore size ratio, and exponentially approaches unity as this ratio decreases [26, 27]. The effects of K_r become small (i.e., the value is close to unity) when the adsorbate size is less than 1/10 of the pore size.

The parameter τ accounts for the deviation of the diffusion path from a straight line. Analogous to Archies law [28], Grathwohl [29] demonstrated that

τ is inversely related to the porosity, n, raised to an exponent, m, as shown in Eq. (66):

$$\tau = 1/n^m \tag{66}$$

For diffusion between soil and sediment particles, n represents the interstitial porosity. For diffusion within soil and sediment particles, n represents the intraparticle porosity. Values of m close to 1 are common for diffusion in porous media [29–31]. However, in low porosity materials this value can increase [32]. A more thorough treatment of this topic can be found in Grathwohl [33].

3.4 Transient Diffusion

3.4.1 Fick's Second Law

Fick's second law defines the behavior of a diffusing chemical in space over time. Fick's second law is derived from Fick's first law and the equation of continuity for a solute. For simplicity, we derive Fick's second law in 1-D coordinates. This can readily be extended to multiple dimensions or to spherical coordinates [22].

First, consider the element of volume in Fig. 4, whose sides are parallel to the axes of coordinates and are of lengths 2dx, 2dy, 2dz.

Let the origin be centered in the box, the concentration of diffusing substance be C [M L^{-3}], and the flux of diffusing substance along the x coordinate be F_x [M L^{-2} t^{-1}]. The rate at which diffusing substance enters the element through the left face is given by

$$4\,dy\,dz\,(F_x - \partial F_x/\partial x\,dx) \tag{67}$$

Similarly, the rate of diffusing substance leaving the element through the right face is given by

$$4\,dy\,dz\,(F_x + \partial F_x/\partial x\,dx) \tag{68}$$

Subtracting the rate of mass out from the rate of mass in yields

$$\text{Rate In} - \text{Rate Out} = -8\,dx\,dy\,dz\,\partial F_x/\partial x \tag{69}$$

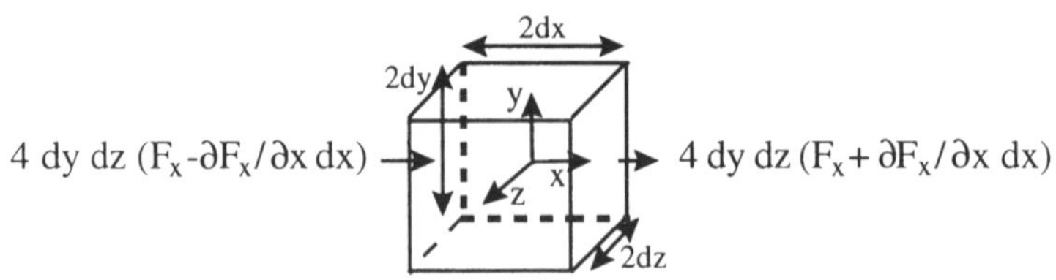

Fig. 4 Representative elementary volume (REV)

The rate of mass accumulation in the box is given by

$$2\ dx\ 2\ dy\ 2\ dz\ \partial C/\partial t = 8\ dx\ dy\ dz\ \partial C/\partial t \tag{70}$$

Performing a mass balance yields

$$\text{Rate of Accumulation} = \text{Rate In} - \text{Rate out} \tag{71}$$

$$\partial C/\partial t = -\partial F_x/\partial x \tag{72}$$

Substituting in for F_x from Fick's first law yields Fick's second law:

$$\partial C/\delta t = -\partial/\partial x\ (-D_{eff}\ \partial C/\partial x) \tag{73}$$

where D_{eff} has been substituted for D to reflect the effects of the porous media on the diffusion coefficient. If the diffusion coefficient is constant (i.e., not a function of concentration or distance):

$$\frac{\partial C}{\partial t} = D_{eff}\frac{\partial^2 C}{\partial x^2} \tag{74}$$

In the natural environment, the diffusion of organic chemicals can be retarded by sorption to soil and sediment grains. To account for this process, Eq. (70) is modified to consider the accumulation of mass in the sorbed phase:

$$\text{Rate of Accumulation} = 8dx\ dy\ dz\ (\partial C/\partial t + (\rho_b/n)\partial q/\partial t) \tag{75}$$

where ρ_b is the bulk density of the porous media. Using the chain rule, the rate of mass accumulation in the sorbed phase can be expressed in terms of the rate of mass accumulation in the aqueous phase as follows:

$$\partial q/\partial t = \partial q/\partial C \cdot \partial C/\partial t \tag{76}$$

At this point any of the isotherms described previously can be used to describe $\partial q/\partial C$. For simplicity, let's assume sorption is linear (i.e., $q=K_dC$). It then follows that

$$\partial q/\partial C = \partial\ (K_dC)/\partial C = K_d \tag{77}$$

Combining Eqs. (75) and (77) gives

$$\text{Rate of Accumulation} = 8\ dx\ dy\ dz\ R\ \partial C/\partial t \tag{78}$$

where $R=(1+\rho_bK_d/n)$ and is defined as the retardation factor. Combining Eqs. (69) and (78) yields

$$\partial C/\partial t = -(1/R)\ \partial F_x/\partial x \tag{79}$$

Using Fick's first law to define F_x gives the form of Fick's second law that accounts for sorption:

$$\frac{\partial C}{\partial t} = \frac{D_{eff}}{R}\frac{\partial^2 C}{\partial x^2} = D_{app}\frac{\partial^2 C}{\partial x^2} \tag{80}$$

where D_{app} is the apparent diffusion coefficient. Equation (80) can be extended to multiple dimensions or other coordinate systems, and R can be defined in terms of other sorption isotherms. For example, in spherical coordinates Eq. (80) is expressed as follows:

$$\frac{\partial C}{\partial t} = D_{app}\left(\frac{\partial^2 C}{\partial r^2} + \frac{2}{r}\frac{\partial C}{\partial r}\right) \tag{81}$$

where r is the radial direction. When $D_{eff}=D_{aq}K_r/\tau$, Eqs. (80) and (81) have been referred to as the pore diffusion model. For a more comprehensive treatment of this topic see Grathwohl [33] or Wu and Gschwend [34].

3.4.2 Solutions of Fick's Second Law

Fick's second law is a partial differential equation that defines the change in concentration within a phase due to the process of molecular diffusion. Fick's second law can be solved numerically, or it can be directly solved to obtain a closed form solution for simplified boundary and initial conditions.

Entire books are devoted to the solution of Fick's second law subject to different boundary and initial conditions, one of the most notable being Crank [22]. Consider the case when a HOP, initially confined to a narrow region between -h and h, is allowed to diffuse away in one dimension to infinity. Formally we can represent these initial and boundary conditions as follows:

$$C(t=0, -h<x<h) = C_0 \tag{82}$$

$$C(t=0, x<-h, x>h) = 0 \tag{83}$$

$$\left.\frac{dC}{dx}\right|_{x=0} = 0 \tag{84}$$

$$C(t, x\rightarrow\infty) = 0 \tag{85}$$

Solving Eq. (80) subject to Eqs. (82)–(85) yields the following solution [22]:

$$C = \frac{1}{2}C_0\left\{\operatorname{erf}\frac{h-x}{2\sqrt{D_{app}t}} + \operatorname{erf}\frac{h+x}{2\sqrt{D_{app}t}}\right\} \tag{86}$$

Equation (86) can be used to approximate diffusion behavior in the environment. For example, consider a small piece or slab of recycled asphalt material (RAM) used in road construction that is 2 h wide and contains the HOP phenanthrene. Just after the RAM has been used in road construction, phenanthrene begins to diffuse away through the surrounding media. If diffusion is dominated in one direction, $D_{app,phenanthrene}=10^{-11}$ $m^2\ s^{-1}$, and h=0.02 m, Eq. (86) can be used to approximate the concentration profiles in the slab. As shown in Fig. 5, mass spreads out in a Gaussian manner with increasing time.

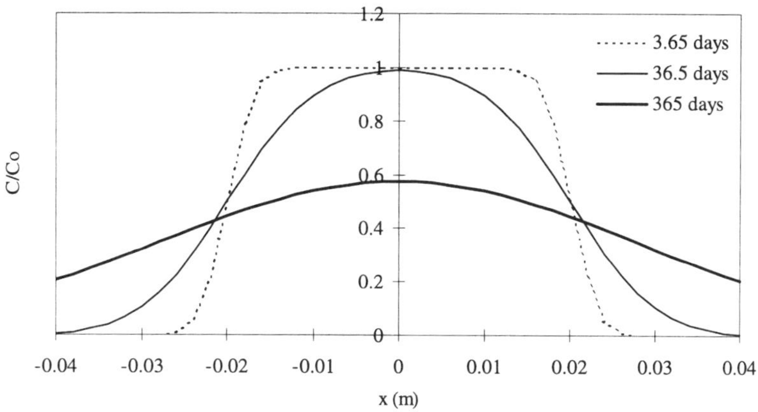

Fig. 5 Profiles for phenanthrene diffusion from a slab

Now consider the case when a HOP diffuses into a sphere of radius a, initially at concentration C_0, and the concentration outside the sphere is constant and equal to C_1. Formally we can represent these boundary and initial conditions as follows:

$$C(t = 0, r < a) = C_0 \tag{87}$$

$$C(t, r = a) = C_1 \tag{88}$$

$$\left.\frac{dc}{dr}\right|_{r=0} = 0 \tag{89}$$

Solving Eq. (81) subject to Eqs. (87)–(89) yields the following solution [22]:

$$\frac{C - C_0}{C_1 - C_0} = 1 + \frac{2a}{\pi r} \sum_{n=1}^{\infty} \frac{(-1)^n}{n} \sin\left(\frac{n\pi r}{a}\right) \exp\left(-D_{app} n^2 \pi^2 t/a^2\right) \tag{90}$$

Equation (90) can be modified to yield the fractional uptake:

$$f = \frac{M_0}{M_\infty} = 1 - \frac{6}{\pi^2} \sum_{n=1}^{\infty} \frac{1}{n^2} \exp\left(D_{app} n^2 \pi^2 t/a^2\right) \tag{91}$$

Equations (90) and (91) can be used to approximate diffusion into porous solids like chunks of asphalt or soil and sediment grains. For example, assume that an HOP is diffusing into a soil grain with $D_{app}=10^{-5}$ $m^2\,s^{-1}$ and $a=10^{-3}$ m. Equation (90) can then be solved to yield the concentration profiles shown in Fig. 6 (where $C_0<C_1$). As shown, the HOP penetrates deeper into the soil grain with increasing time.

Instead of a linear isotherm, a nonlinear isotherm could also have been used to define $\partial q/\partial C$ in Eq. (77). For example, if the Freundlich isotherm were used the retardation factor is defined as follows:

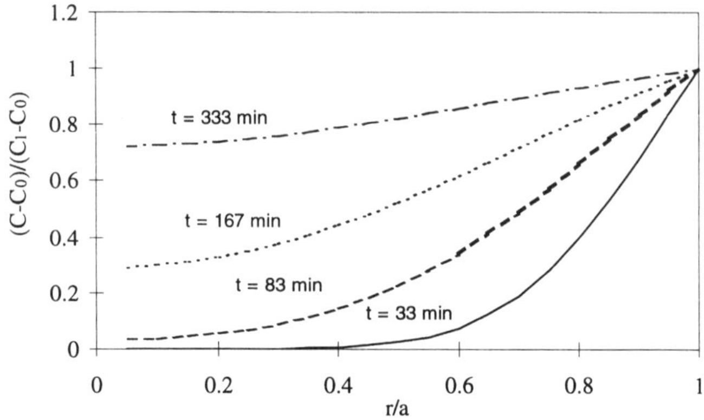

Fig. 6 Profiles for a HOP diffusing into a sphere

$$R = 1 + \frac{\rho_b}{n} n_F K_F C^{n_F - 1} \tag{92}$$

When sorption is nonlinear D_{app} is a function of concentration. This makes solving Fick's second law (Eq. 80 or 81) considerably more complex because no analytical solutions are available. Instead, a numerical solution is required. This also makes the solution for Fick's second law more difficult to interpret. For example, if $n_F<1$ the rate of diffusion is slower at lower concentrations than at higher concentrations.

Results from an explicit in time finite different solution for Fick's second law in radial coordinates subject to the boundary and initial conditions in Eqs. (87)–(89) (where $C_0>C_1$ and $C_1=0$) is shown in Fig. 7. At late time mass leaves the sphere much more slowly when $n_F<1$.

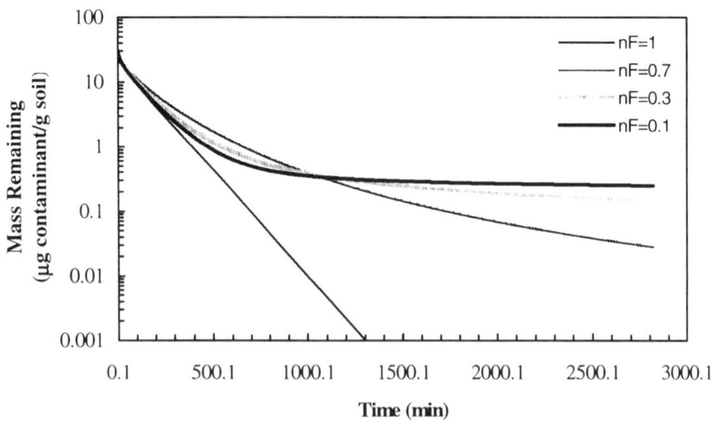

Fig. 7 Profiles of a HOP diffusing from a sphere subject to nonlinear sorption (D_{eff}=4.0 ×10^{-7} cm^2/s, K_F=1 (µg/g)(µg/ml)^{nF}, r=0.16 cm, ρ_{grain}=1 g/ml, n_{grain}=0.15)

3.4.3 Diffusion in Natural Solids

Sorption and desorption of organic chemicals in natural solids are generally considered diffusion controlled processes. A typical desorption profile for a VOC is shown in Fig. 8. Here nitrogen is purged through a column that contains trichloroethene (TCE) sorbed to Santa Clara sediment at 100% relative humidity. Prior to purging, TCE was allowed to equilibrate with the Santa Clara sediment for over one month. As shown, over 60% of the TCE in the column is removed within minutes (this includes both mass in the vapor phase between grains and sorbed mass), while the remaining mass takes weeks to months to desorb. This type of behavior has led researchers to conclude that there are both fast and slow desorbing fractions of mass. At least some of this slow behavior can be attributed to nonlinear sorption. Figure 7 illustrated how concentration profiles tail at long times when $n_F<1$.

Predicting fast and slow rates of sorption and desorption in natural solids is a subject of much research and debate. Often times fast sorption and desorption are approximated by assuming equilibrium portioning between the solid and the pore water, and slow sorption and desorption are approximated with a diffusion equation. Such models are often referred to as dual-mode models and several different variants are possible [35–39]. Other times two diffusion equations were used to approximate fast and slow rates of sorption and desorption [31, 36]. For example, for a VOC Werth and Reinhard [31] used the pore diffusion model to predict fast desorption, and a separate diffusion equation to fit slow desorption. Fast and slow rates of sorption and desorption have also been modeled using one or more distributions of diffusion rates (i.e., a superposition of solutions from many diffusion equations, each with a different diffusion coefficient) [40–42].

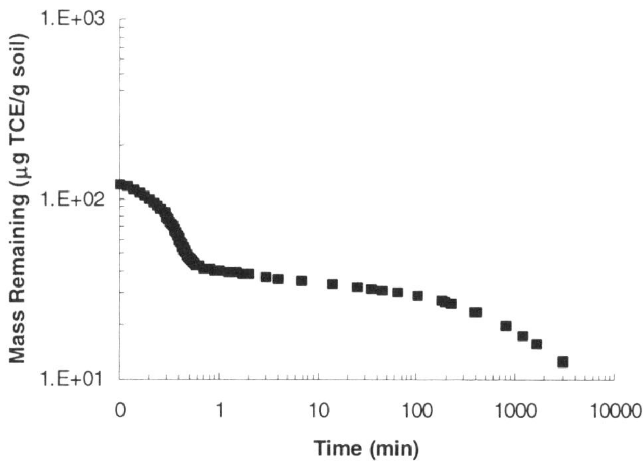

Fig. 8 Desorption profiles for trichloroethene (TCE) from Santa Clara sediment

Depending on the distribution chosen, as few as three fitting parameters may be required to define a distribution of diffusion rates. In some cases, a single distribution was used to describe both fast and slow rates of sorption and desorption, and in other cases fast and slow mass transfer were captured with separate distributions of diffusion rates. For example, Werth et al. [42] used the pore diffusion model with nonlinear sorption to predict fast desorption, and a gamma distribution of diffusion rate constants to describe slow desorption.

The first order equation has also been used to describe sorption and desorption processes. Haggerty and Gorelick [43] showed that there is an equivalent distribution of first order rate constants for any distribution of diffusion coefficients. Other models that use distributions of first order rate constants are presented in the following references [44–47].

3.5 Mass Transfer Between Phases

The rate and extent that a HOP transfers between phases is determined by the fugacity gradient. For example, when the fugacity in water is greater than the fugacity in air, the HOP is transferred from water to air. Similarly, when the fugacity in a solid is greater than the fugacity in water, the HOP is transferred from the solid to water.

Several theories have been developed to describe the rate of interphase mass transfer. These include film theory, boundary layer theory, penetration theory, and surface renewal theory. In this chapter we will review the first two, along with an overview of empirical correlations that are used to describe mass transfer. A more thorough overview of mass transfer theories can be found in Bird, Stewart and Lightfoot [48], Clark [49], Logan [50], and Weber and DiGiano [51].

3.5.1 Film Theory

Film theory is the oldest approach for describing interphase mass transfer. While its assumptions are often not valid (especially for laminar flow in porous media), it provides a conceptual picture that is useful when considering interphase mass transfer. Film theory was first proposed by Lewis [52] and Whitman [53]. The theory is based on the assumptions that (1) mass transfer from a turbulent fluid phase can be treated by assuming that the entire resistance to mass transfer resides in a stagnant film at the phase interface, (2) equilibrium obtains at the interface, (3) the bulk fluid phase is sufficiently well-mixed so that concentration gradients in the bulk fluid are negligible, and (4) the concentration gradient in the stagnant boundary layer (or film) is linear. These assumptions are illustrated in Fig. 9 for an air-water system.

Since mass transfer is driven by the fugacity gradient, we can use the water/air fugacity relationship to convert the gas phase concentrations to equivalent liquid-phase concentrations as shown in Fig. 10.

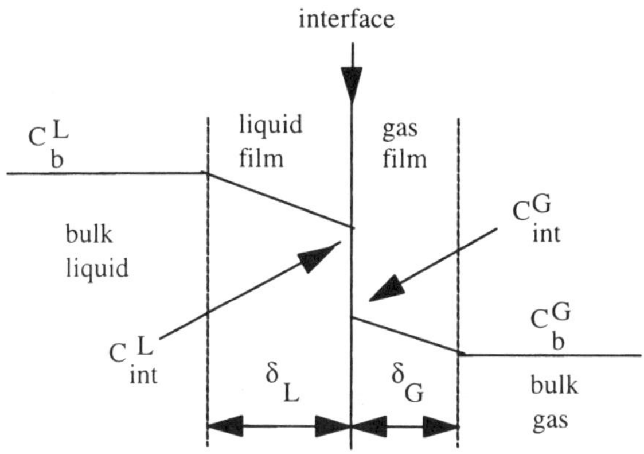

Fig. 9 Illustration of film theory assumptions

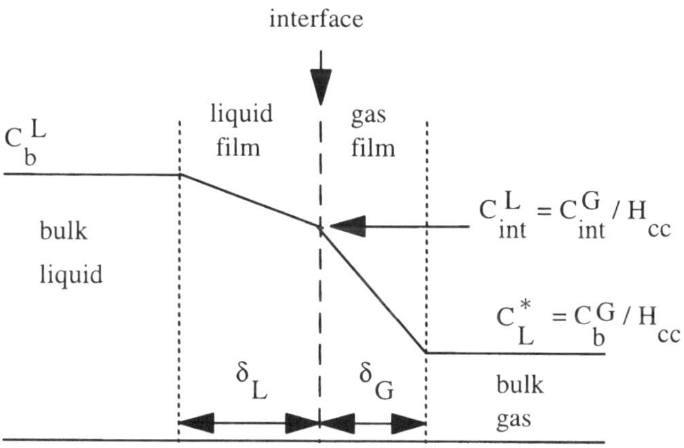

Fig. 10 Illustration of film theory assumptions with all concentrations expressed with respect to the liquid phase

It follows that the mass flux per unit area is defined by an expression analogous to Fick's first law as follows:

$$F = (1/A)\, dM/dt = -(D_L/\delta_L)\,(C_b^L - C_{int}^L) = -(D_G/\delta_G)(C_{int}^L - C_L^*) \tag{93}$$

where M is the mass, A is the area across which mass transfer occurs $[L^2]$, t is the time, D_L is the liquid phase diffusivity $[L^2\, t^{-1}]$, D_G is the gas phase diffusivity $[L^2\, t^{-1}]$, δ_L is the thickness of the liquid phase stagnant zone [L], δ_G is the thickness of the gas phase stagnant zone [L], D_L/δ_L is the specific liquid phase mass transfer coefficient (k_L,$[L\, t^{-1}]$), D_G/δ_G is the specific gas phase mass transfer coefficient (k_G,$[L\, t^{-1}]$), C_b^L is the bulk concentration in the liquid, C_{int}^L is the liquid interface concentration, and C_L^* is the liquid phase equivalent gas phase concentration. Equation (93) can be expressed as the mass flux as follows:

$$dM/dt = -k_L\,A\,(C_b^L - C_{int}^L) = -k_G\,A\,(C_{int}^L - C_L^*) \qquad (94)$$

Because the concentration at the interface is difficult to determine, Eq. (94) is modified to define the mass transfer rate between the two phases in terms of bulk phase concentrations and an overall mass transfer coefficient:

$$dM/dt = -K_L\,A\,(C_b^L - C_L^*) = -K_G\,A\,(C_G^* - C_b^G) \qquad (95)$$

where K_L is the overall mass transfer coefficient [L t^{-1}] based on liquid concentrations, K_G is the overall mass transfer coefficient [L t^{-1}] based on gas phase concentrations, C_G^* is the equivalent gas phase concentration of the bulk liquid, and C_b^G is the bulk phase gas concentration.

Film theory predicts that the mass transfer coefficient for a phase (or the overall mass transfer coefficient) is proportional to the diffusion coefficient and inversely proportional to the thickness of the stagnant zone. The diffusion coefficient can be calculated from either the Wilke-Chang or the FSG equations. However, δ is difficult (if not impossible) to determine. Hence, mass transfer coefficients are often determined from empirical correlations. Also, Film theory is based on the assumption that the bulk fluid phases are perfectly mixed. While this might approach reality for well-mixed turbulent systems, this is certainly not the case for laminar systems.

3.5.2
Boundary Layer Theory

The boundary layer theory is based on the system in Fig. 11, where mass transfer is occurring from a flat plate in the presence of laminar flow.

The plate represents an infinite source of mass, such that the concentration at the plate surface is constant. The fluid represents a mass sink. Flow over the flat plate is uniform and parallel to the surface. Hence, the x component of the Navier-Stokes equation and the continuity equation in two dimensions (x and y) describe flow over the plate. These equations are given as

$$u_x \frac{\partial u_x}{\partial x} + u_x \frac{\partial u_y}{\partial y} = -\frac{\partial p}{\partial x} + \nu \left(\frac{\partial^2 u_x}{\partial x^2} + \frac{\partial^2 u_x}{\partial y^2} \right) \qquad (96)$$

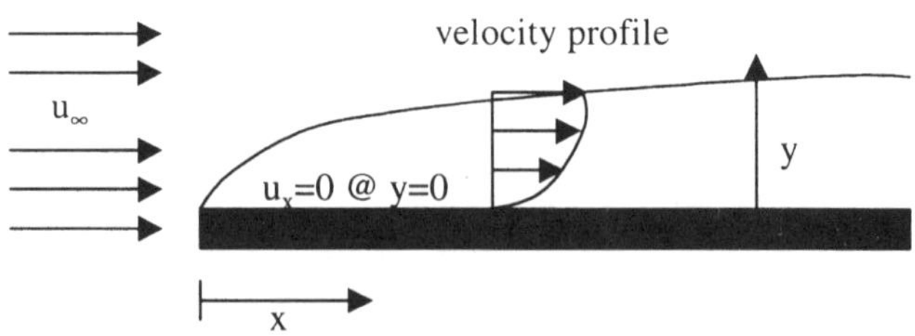

Fig. 11 Velocity profile over a flat plate

$$\frac{\partial u_x}{\partial x} + \frac{\partial u_y}{\partial y} = 0 \tag{97}$$

where p is the pressure, ν is the kinematic viscosity [$L^2\ t^{-1}$], and u is the velocity [$L\ t^{-1}$]. If we assume that the boundary layer is very thin (i.e., $\partial p/\partial x=0$ and $\partial^2 u/\partial y^2 \gg \partial^2 u/\partial x^2$), then the Navier-Stokes equation can be simplified to

$$u_x \frac{\partial u_x}{\partial x} + u_x \frac{\partial u_y}{\partial y} = \nu \left(\frac{\partial^2 u_x}{\partial y^2} \right) \tag{98}$$

The appropriate boundary conditions for flow over the flat plate are $u_x=u_y=0$ at $y=0$ and $u_x=u_\infty$ at $y=\infty$. Given these, Eqs. (96) and (98) can be solved to determine the steady state velocity profile adjacent to the flat plate. If we assume that the concentration and velocity profiles have similar shape, the mass flux per unit area from the surface of the plate is defined as follows [50]:

$$F_{C,y=0} = \frac{0.332\,D}{x}\ Re^{1/2} Sc^{1/3} (C_{y=0} - C_{y=\infty}) \tag{99}$$

where D is the diffusion coefficient in the fluid above the plate, Re is the Reynolds number, and Sc is the Schmidt number. Re and Sc are defined as

$$Re = u_\infty x/\nu \tag{100}$$

$$Sc = \nu/D \tag{101}$$

Re represents the ratio of advective to viscous forces. Sc represents the ratio of viscous to diffusive forces. Sc accounts for differences in the thickness of the velocity and concentration boundary layers. For air $Sc \sim 1$, while for water $Sc \gg 1$.

Equation (99) is very similar in form to Eq. (93). In fact, these equations have the same form if we make the following substitution:

$$k = \frac{0.332\,D}{x}\ Re^{1/2} Sc^{1/3} \tag{102}$$

where k is a generic mass transfer rate coefficient. Equations (99) and (102) show that mass transfer from a flat plate can be represented with a form of Fick's first law (often called the linear driving force model), where the mass transfer rate coefficient is a function of Re and Sc. The mass transfer rate coefficient is frequently defined in terms of the dimensionless Sherwood number as follows:

$$Sh = \frac{kx}{D} = 0.332\ Re^{1/2} Sc^{1/3} \tag{103}$$

While Eq. (103) was obtained from first principles, the approach of quantifying mass transfer with a mass transfer rate coefficient obtained from the linear driving force model, and expressing the mass transfer rate coefficient in terms of other dimensionless numbers, is the basis for many empirical models used to describe mass transfer. This is further discussed in the next section.

3.5.3
Empirical Correlations

The linear driving force model (a form of Fick's first law) is often used to determine interphase mass transfer rate coefficients. This equation is expressed as

$$F = k(C_s - C_\infty) \tag{104}$$

where F is the mass flux per unit area $[M\ L^{-2}\ t^{-1}]$, k is the mass transfer rate coefficient $[L\ t^{-1}]$, C_s is the concentration at the surface where mass transfer occurs, and C_∞ is the concentration far away from the mass transfer surface. The parameter k is typically expressed as the dimensionless Sherwood number (Sh) as follows:

$$Sh = kl/D \tag{105}$$

where l is the characteristic length scale. Sh is then determined as a function of different experimental variables and relevant correlations are developed. For example, Froessling obtained the following relationship for chemical transport into gas bubbles [50]:

$$Sh = 1 + 0.39\ Re^{1/2} Sc^{1/3} \tag{106}$$

Empirical relationships of similar form for mass transfer in packed beds, to falling spheres, to flat plates, and other relevant geometries are given in several excellent text books [48–51].

Another mass transfer process that has received considerable attention in recent years is the dissolution of nonaqueous phase liquids (NAPLs), such as grease, oils, fuels, and solvents. Such liquids may be found in construction and repair materials, and are thus of interest. Consider NAPL trapped as isolated blobs or ganglia in saturated porous media. The governing equation for dissolution is given as

$$q\frac{dC}{dx} = ka(C_s - C) \tag{107}$$

where q is the Darcy velocity and a is the specific NAPL-water interfacial area (NAPL-water interfacial area divided by the system volume $[L^2\ L^{-3}]$). The parameter a is difficult, if not impossible, to measure in real systems. Hence, k and a are often lumped into a bulk mass transfer coefficient (K=ka) and defined by a modified Sherwood number, Sh'. This latter parameter is defined as

$$Sh' = K\ l^2/D \tag{108}$$

Several published correlations for Sh′ are given in Table 5. Presently, scientists and engineers are trying to determine viable surrogates for a, such that Sh′ can be expressed in terms of measurable system parameters, but still be sensitive to changes in the parameter a during dissolution [54, 55].

Table 5 Sherwood number correlations for NAPL dissolution

Reference	Correlation	Condition
56	$Sh' = 12\,Re^{0.75}\,\theta_n^{0.6}\,Sc^{0.5}$	Steady state, glass bead column, $0.016 \le \theta_n \le 0.07$, $0.0015 \le Re \le 0.1$
57	$Sh' = 57.7\,Re^{0.61}\,d_{50}^{0.64}\,U_i^{0.41}$	Steady state sand column $0.032 \le \theta_n \le 0.064$, $0.012 \le Re \le 0.2$
58	$Sh' = 150\,\theta_n^{0.79}\,Re^{0.87}$	Steady state sand column, $0 \le \theta_n \le 0.04$, $0.00012 \le Re \le 0.021$ and
	[a] $Sh' = 340\,\theta_n^{0.87}\,Re^{0.71}\left(\frac{x}{\overline{d_p}}\right)^{-0.31}$	$1.4 \le \frac{x}{\overline{d_p}} \le 180$
59	[b] $Sh' = 4.13\,Re'^{0.598}\frac{d_{50}^{0.673}}{d_M}\,U_i^{0.369}\left(\frac{\theta_n}{\theta_{n,0}}\right)^{\beta_4}$ where $\beta_4 = 0.518 + 0.114\frac{d_{50}}{d_M} + 0.1\,U_i$	Transient sand column, $0.039 \le \theta_n \le 0.065$, $0.06 \le Re' \le 1.33$

[a] x=the distance into the region of residual NAPL [L], $\overline{d_p}$=mean grain diameter [L].
[b] $Re'=Re/n$, d_M=medium graindiameter [L], θ_n=volumetric NAPL content [-], $\theta_{n,0}$=initial volumetric NAPL content [-], U_i=grain uniformity index [-], d_{50}=median grain diameter [L].

4 Conclusions

Equilibrium partitioning and mass transfer relationships that control the fate of HOPs in CRM and in different phases in the environment were presented in this chapter. Partitioning relationships were derived from thermodynamic principles for air, liquid, and solid phases, and they were used to determine the driving force for mass transfer. Diffusion coefficients were examined and those in water were much greater than those in air. Mass transfer relationships were developed for both transport within phases, and transport between phases. Several analytical solutions for mass transfer were examined and applied to relevant problems using calculated diffusion coefficients or mass transfer rate constants obtained from the literature. The equations and approaches used in this chapter can be used to evaluate partitioning and transport of HOP in CRM and the environment.

Acknowledgements Corrections to the text from Jun Li at the Connecticut Agricultural Experiment Station are acknowledged.

References

1. Cooper AR (1997) Cooper's toxic exposures desk reference with CD-ROM. CRC Press, Boca Raton, Florida
2. Brantley AS, Townsend TG (1999) Leaching of pollutants from reclaimed asphalt pavement. Environ Eng Sci 16:105–116
3. Lewis GN (1901) Das Gesetz physiko-chemischer, Z Physik Chem 38:205–226
4. Mackay D, Paterson S (1981) Calculating fugacity. Environ Sci Technol 15:1006–1014
5. Ashworth RA, Howe GB, Mullins ME, Rogers TN (1988) Air-water partitioning coefficients of organics in dilute aqueous solutions, J Hazard Mater 18:25–36
6. Reinhard M, Drefahl A (1999) Handbook for estimating physicochemical properties of organic compounds. Wiley, New York
7. Karickhoff SW, Brown DS, Scott TA (1979) Sorption of hydrophobic pollutants on natural sediments. Water Res 13:241–248
8. Grathwohl P (1990) Influence of organic matter from soils and sediments from various origins on the sorption of some chlorinated aliphatic hydrocarbons: Implications on Koc correlations. Environ Sci Technol 24:1687–1693
9. Ruthven DM (1984) Principles of adsorption and adsorption processes. Wiley, New York
10. Xia G, Ball WP (1999) Adsorption-partitioning uptake of nine low-polarity organic chemicals on a natural sorbent. Environ Sci Technol 33:262–269
11. Xing B, Pignatello JJ, Gigliotti B (1996) Competitive sorption between atrazine and other organic compounds in soils and model sorbents. Environ Sci Technol 30:2432–2440
12. LeBoeuf EJ, Weber WJ Jr (1997) A distributed reactivity model for sorption by soils and sediments. 8. Sorbent organic domains: discovery of a humic acid glass transition and an argument for a polymer-based model, Environ Sci Technol 31:1697–1702
13. Xing B, Pignatello JJ (1997) Dual-mode sorption of low-polarity compounds in glassy poly(vinyl-chloride) and soil organic matter. Environ Sci Technol 31:792–799
14. Chiou CT, Kile DE (1998) Deviations from sorption linearity on soils of polar and non-polar organic compounds at low relative concentrations. Environ Sci Technol 32:338–343
15. Kleineidam S, Rugner H, Ligouis B, Grathwohl P (1999) Organic matter facies and equilibrium sorption of phenanthrene. Environ Sci Technol 33:1637–1644
16. Gustafsson O, Haghseta F, Chan C, MacFarlane J, Gschwend PM (1997) Quantification of the dilute sedimentary soot phase: implications for PAH speciation and bioavailability. Environ Sci Technol 31:203–209
17. Li J, Werth CJ (2001) Evaluating competitive sorption mechanisms of volatile organic compounds in soils and sediments using polymers and zeolites. Environ Sci Technol 35:568–574
18. Lide DR (ed) (1991) CRC handbook of chemistry and physics, 72nd edn. CRC Press, Ann Arbor, MI
19. Chiou CT, McGroddy SE, Kile DE (1998) Partition characteristics of polycyclic aromatic hydrocarbons on soils and sediments. Environ Sci Technol 32:264–269
20. Freeze RA, Cherry JA (1979) Groundwater. Prentice-Hall, Englewood Cliffs, NJ
21. Atkins PW (1990) Physical chemistry. Freeman, New York
22. Crank J (1975) The mathematics of diffusion, 2nd edn. Oxford University Press, New York
23. Wilke, Chang (1955) Correlation of diffusion coefficients in dilute solutions, AIChE J 1:264–270
24. Reid RC, Prauznitz JM, Sherwood TK (1977) The properties of gases and liquids, 3rd edn. McGraw-Hill, New York
25. Lyman WJ, Reehl WF, Rosenblatt DH (1982) Handbook of chemical property estimation methods. American Chemical Society, Washington, DC

26. Chantong A, Massoth FE (1983) Restrictive diffusion in aluminas. AIChE J 29:725–731
27. Satterfield CN, Colton CK (1973) Restricted diffusion in liquids within fine pores. AIChE J 19:628–635
28. Archie GE (1942) The electrical resistivity log as an aid in determining some reservoir characteristics. Trans AIME 146:54–62
29. Grathwohl P (1992) Diffusion controlled desorption of organic contaminants in various soils and rocks. Seventh International Symposium on Water Rock Interactions, Park City, Utah
30. Wakao N, Smith JM (1962) Diffusion in catalyst pellets, Chem Eng Sci 17:825–834
31. Werth CJ, Reinhard M (1997) Effects of temperature on trichloroethylene desorption from silica gel and natural sediments. 2. Kinetics. Environ Sci Technol 31:697–703
32. Wong PZ, Koplik J, Tomanic JP (1984) Conductivity and permeability of rocks. Phys Rev B 30(11):6606–6614
33. Grathwohl P (1998) Diffusion in natural porous media, contaminant transport, sorption/desorption and dissolution kinetics. Kluwer Academic, Boston
34. Wu SC, Gschwend PM (1986) Sorption kinetics of hydrophobic organic compounds to natural sediments and soils. Environ Sci Technol 20:717–725
35. Ball WP, Roberts PV (1991) Long-term sorption of halogenated organic chemicals by aquifer material. 2. Intraparticle diffusion. Environ Sci Techol 25:1237–1249
36. Brusseau ML, Srivastava, R (1999) Nonideal transport of reactive solutes in heterogeneous porous media. 4. Analysis of the Cape Cod natural-gradient field experiment. Water Resour Res 35:1113–1125
37. Cameron DL, Klute A (1997) Convective-dispersive solute transport with a combined equilibrium and kinetic adsorption model. Water Resour Res 13:183–188
38. Coats KH, Smith BD (1964) Dead-end pore volume and dispersion in porous media. Soc Pet Eng J 4:73–84
39. van Genuchten MT, Wierenga PJ (1976) Mass transfer studies in sorbing porous media. I. Analytical solutions. Soil Sci Soc Am J 40:473–480
40. Cunningham JA, Werth CJ, Reinhard M, Roberts PV (1997) Effects of grain-scale mass transfer on the transport of volatile organics through sediments. 1. Model development. Water Resour Res 33(12):2713–2726
41. Pedit JA, Miller CT (1994) Heterogenous sorption processes in subsurface systems. 1. Model formulation and applications. Environ Sci Technol 28:2094–2104
42. Werth CJ, Cunningham JA, Roberts PV, Reinhard M (1997) Effects of grain-scale mass transfer on the transport of volatile organics through sediments. 2. Column results. Water Resour Res 33:2727–2740
43. Haggerty R, Gorelick SM (1998) Modeling mass transfer processes in soil columns with pore-scale heterogeneity. Soil Sci Soc Am J 62:62–71
44. Chen W, Wagenet RJ (1995) Solute transport in porous media with sorption site heterogeneity. Environ Sci Technol 29:2725–2734
45. Culver TB, Hallisey SP, Sahoo D, Deitsch JJ, Smith JA (1997) Modeling the desorption of organic contaminants from long-term contaminated soil using distributed mass transfer rates. Environ Sci Technol 31:1581–1588
46. Culver TB, Brown RA, Smith JA (2000) Rate-limited sorption and desorption of 1,2-dichlorobenzene to a natural sand soil column. Environ Sci Technol 34:2446–2452
47. Lorden SW, Chen W, Lion LW (1998) Experiments and modeling of the transport of trichloroethene vapor in unsaturated aquifer material. Environ Sci Technol 32:2009–2017
48. Bird RB, Stewart WE, Lightfoot EN (1960) Transport phenomena. Wiley, New York
49. Clark MM (1996) Transport modeling for environmental engineers and scientists. Wiley, New York

50. Logan BE (1999) Environmental transport processes. Wiley, New York
51. Weber WJ Jr, DiGiano FA (1996) Process dynamics in environmental systems. Wiley, New York
52. Lewis WK (1916) The principles of counter-current extraction. J Ind Eng Chem 8:825–833
53. Whitman WG (1923) Preliminary experimental confirmation of the two-film theory of gas adsorption. Chem Met Eng 29:146–148
54. Johns ML, Gladden LF (1999) Magnetic resonance imaging study of the dissolution kinetics of octanol in porous media. J Colloid Int Sci 210:261–270
55. Zhang C, Werth CJ, Webb AG (2002) A magnetic resonance imaging study of dense non-aqueous phase liquid dissolution from angular porous media. Environ Sci Technol 36, 3310–3317
56. Miller CT, Poirier-McNeil MM, Mayer AS (1990) Dissolution of trapped nonaqueous phase liquids: mass transfer characteristics. Water Resour Res 26:2783–2796
57. Powers SE, Abriola LM, Weber WJ Jr (1992) An experimental investigation of nonaqueous phase liquid dissolution in saturated subsurface system: steady state mass transfer rates. Water Resour Res 28:2691–2705
58. Imhoff PT, Jaffe' PR, Pinder GF (1993) An experiment study of complete dissolution of a nonaqueous phase liquid in saturated porous media. Water Resour Res 30:307–320
59. Powers SE, Abriola LM, Weber WJ Jr (1994) An experimental investigation of nanaqueous phase liquid dissolution in saturated subsurface systems: transient mass transfer rates. Water Resour Res 30:321–332

Handb Environ Chem Vol. 5, Part F, Vol. 3 (2005): 33–63
DOI 10.1007/b11438

Organic Chemicals in Groundwater: Modeling Fate and Transport

Mark N. Goltz[1] (✉) · Jae-Woo Park[2] · Peter P. Feng[3] · Harold C. Young[4]

[1] Department of Systems and Engineering Management, Graduate School of Engineering and Management, Air Force Institute of Technology, 2950 Hobson Way, Bldg 640, Wright-Patterson Air Force Base, OH 45433-7765, USA
mark.goltz@afit.edu

[2] Department of Civil Engineering, Hanyang University, 17 Haengdang-dong, Seongdong-gu, Seoul, 133-791, South Korea
jaewoopark@hanyang.ac.kr

[3] PSC2 Box 11694, APO AE 09012
peter.feng-02@ramstein.af.mil

[4] 1st FSSG MARFORPAC, Box 555606 G-4/IH DIV B-14172, Camp Pendleton, CA92055-5606, USA
YoungHC@1FSSG.USMC.MIL

1 Introduction 35

2 Fate and Transport Processes 35
2.1 Physicochemical Processes 36
2.1.1 Advection 36
2.1.2 Dispersion 36
2.1.3 Sorption 37
2.2 Transformation Processes 40
2.2.1 Hydrolysis 41
2.2.2 Dehydrohalogenation 41
2.2.3 Oxidation Reactions 42
2.2.4 Reduction Reactions 46

3 Modeling 48
3.1 General Advective/Dispersive Reactive Transport Equation 49
3.2 Model Solutions to the Transport Equation 50
3.2.1 Analytical 50
3.2.2 Numerical 51
3.2.3 Model Codes 51
3.3 Case Studies of Model Applications 53
3.3.1 Modeling Petroleum Hydrocarbon Fate and Transport During a Field Experiment at Canadian Forces Base (CFB) Borden 54
3.3.2 Modeling Petroleum Hydrocarbon Fate and Transport at a Contaminated Site on Patrick Air Force Base 55
3.3.3 Modeling Chlorinated Ethene Fate and Transport at a Contaminated Site on Dover Air Force Base 55
3.3.4 Modeling Chlorinated Ethene Fate and Transport at a Contaminated Site on F.E. Warren Air Force Base 56
3.3.5 Modeling Petroleum Hydrocarbon and Chlorinated Ethene Fate and Transport at a Contaminated Site on Plattsburgh Air Force Base 56

4 Discussion 58
4.1 Extending Laboratory Knowledge to the Field 58
4.2 Understanding Basic Processes 59

References 59

Abstract Hazardous organic chemical waste that has been placed in landfills or recycled for use as highway construction and repair materials has the potential of leaching through the vadose zone into the groundwater. The groundwater may then serve as a pathway to transport these chemical contaminants to human and environmental receptors. Thus, it is important that we understand how processes that occur in groundwater affect the fate and transport of these organic chemicals. Groundwater fate and transport models are available to help us to gain this understanding, as well as to quantify the concentrations at which these organic chemicals reach receptors, so that we may be able to quantify risk.

In this chapter, we first present the physical, chemical, and biological processes that affect organic chemical fate and transport in groundwater. We also present models of these individual processes. In the second part of the chapter, we examine models that combine the individual processes, so as to comprehensively simulate how organic chemical concentrations in a contaminated groundwater system vary over space and time. Finally, we present case studies that demonstrate the utility of fate and transport modeling in helping us to understand the behavior of organic contaminants in groundwater.

Keywords Organic contamination · Groundwater · Fate and transport modeling · Biodegradation · Chemical transformation · Advection · Dispersion · Sorption · Analytical model · Numerical model

List of Symbols and Abbreviations

BTEX	Benzene, toluene, ethylbenzene, and xylene isomers
C	Dissolved contaminant concentration in the groundwater $[ML^{-3}]$
CF	Trichloromethane (chloroform)
CT	Tetrachloromethane (carbon tetrachloride)
D_i	Dispersion coefficient in the i-th direction $[L^2T^{-1}]$
$\approx{D}$	Dispersion coefficient matrix $[L^2T^{-1}]$
DCA	Dichloroethane
DBM	Dibromomethane
DCE	Dichloroethylene
ETH	Ethene
h	Hydraulic head of groundwater at a location in space [L]
∇h	Hydraulic gradient [-]
HCA	Hexchloroethane
$\approx{K}$	Hydraulic conductivity tensor of the porous medium $[LT^{-1}]$
K_d	Sorption partitioning or distribution coefficient $[L^3M^{-1}]$
K_f	Freundlich sorption coefficient $[M^{-N}L^{3N}]$
K	Langmuir sorption coefficient $[L^3M^{-1}]$
k_f	Sorption rate constant $[L^3M^{-1}T^{-1}]$
k_b	First-order desorption rate constant $[T^{-1}]$
k	First-order degradation rate constant $[T^{-1}]$
k_{max}	Maximum rate of transformation in the Monod expression $[ML^{-3}T^{-1}]$
K_M	Half-saturation constant in the Monod expression $[ML^{-3}]$

K_I Inhibition concentration of a self-inhibitory compound (Haldane kinetics) $[ML^{-3}]$
n Porosity of aquifer material [-]
N Freundlich exponent [-]
PCA Pentachloroethane
PCE Tetrachloroethylene
Q^0 Maximum amount of contaminant that can be sorbed by the solid $[MM^{-1}]$
r Sink due to chemical/biological reaction $[ML^{-3}T^{-1}]$
S Sorbed concentration [M contaminant per M solids]
TBM Tribromomethane
TCA Trichloroethane
TCE Trichloroethylene
TeCA Tetrachloroethane
VC Vinyl chloride
v_x Average linear velocity of groundwater in the x-direction $[LT^{-1}]$
$\tilde{v}$ Groundwater average linear velocity vector $[LT^{-1}]$
α_i Dispersivity in the i-th-direction [L]
ρ_b Bulk density of aquifer solids $[ML^{-3}]$

1 Introduction

Hazardous organic chemical waste that has been placed in landfills or recycled for use as highway construction and repair materials has the potential of leaching through the vadose zone into the groundwater. The groundwater may then serve as a pathway to transport these chemical contaminants to human and environmental receptors. Thus, in order to assess risk, it is important that we understand, and be able to quantitatively predict, the impact of the various physical, chemical, and biological processes that occur in the groundwater upon the fate and transport of these contaminants. In the discussion below, we look at how these processes affect organic chemical fate and transport in the groundwater, and how we can use mathematical models of the processes to quantify the potential for the contaminants to reach receptors.

2 Fate and Transport Processes

For convenience, we will categorize processes occurring in the groundwater as either being physicochemical or transformation processes. Of course, in reality, physicochemical and transformation processes act upon contaminants in concert. We will restrict our discussion to transport of dissolved contaminants in a porous media, as it is typically contaminants dissolved in groundwater that have the greatest potential to move from a source area to downgradient receptors. A detailed discussion of the dissolution of separate phase contaminant in the source area is reported by Chrysikopoulos [1].

2.1 Physicochemical Processes

The physicochemical processes of advection, dispersion, and sorption play a crucial role in the fate and transport of contaminants.

2.1.1 Advection

Advection is the transport of dissolved contaminant mass due to the bulk flow of groundwater, and is "by far the most dominant mass transport process" [2]. Thus, if one understands the groundwater flow system, one can predict how advection will transport dissolved contaminant mass. The speed and direction of groundwater flow may be characterized by the average linear velocity vector ($\tilde{v}$). The average linear velocity of a fluid flowing in a porous medium is determined using Darcy's Law [2]:

$$\tilde{v} = -\frac{\approx{K}}{n}\nabla h \tag{1}$$

where $\approx{K}$ is the hydraulic conductivity tensor of the porous medium [LT^{-1}], h is the hydraulic head of the groundwater at a location in space [L], ∇h is the hydraulic gradient [-], and n is the porosity of aquifer material [-].

Porosity and hydraulic conductivity are both properties of the porous medium. Hydraulic heads, and the hydraulic head gradient, may be determined by solving the equations of flow, with appropriate initial and boundary conditions [2].

2.1.2 Dispersion

Dispersion is the spreading of mass beyond the region that would be affected by advection alone. There are two processes that lead to dispersion, diffusion and mechanical dispersion [2]. Molecular diffusion is the movement of molecules from areas of high concentration to areas of low concentration. Molecular diffusion is only important at extremely low flow velocities and is usually ignored when modeling contaminant transport from source areas to receptors [3]. Mechanical dispersion represents the spreading of contaminant molecules due to spatial variation of groundwater flow velocity. That is, contaminant molecules being carried along with the flow by advection do not move at the groundwater average linear velocity. In reality, the individual contaminant molecules flow faster and slower than the average linear velocity due, for instance, to the tortuosity of the porous medium and the parabolic distribution of groundwater velocities within pores. Mechanical dispersion also accounts for the heterogeneity of the porous medium, where some groundwater may be flowing relatively fast through high hydraulic conductivity regions and some relatively slowly through low hydraulic conductivity regions.

Mechanical dispersion has been shown to be proportional to the average linear velocity, and can be modeled using the following equation, where the coordinate axes have been aligned so that groundwater velocity is in the x-direction [3]:

$$D_i = \alpha_i v_x \tag{2}$$

where D_i is the dispersion coefficient in the i-th direction $[L^2T^{-1}]$, α_i is the dispersivity in the i-th-direction [L], and v_x is the average linear velocity of groundwater in the x-direction $[LT^{-1}]$.

Dispersivity is a property of the porous medium that has been shown to be proportional to the scale of the system under consideration [4]. Thus, the dispersivity of a porous medium in a laboratory column will be considerably smaller than the dispersivity of an aquifer through which contaminated groundwater is flowing over distances of hundreds of meters. Presumably, this scale-effect is due to the increased spreading caused by variations in velocities due to larger scale heterogeneities [2].

2.1.3
Sorption

Sorption is the process where a sorbate binds to a sorbent. In this discussion, the sorbate is the organic contaminant and the sorbent is the aquifer solid. Sorption has a large impact on the rate at which a contaminant is transported, as well as on its ultimate fate. Since sorbed contaminant is immobile, the sorption process retards contaminant transport. Also, since sorbed contaminant is frequently assumed to be unavailable for microbial degradation [5–7], sorption slows the rate at which a contaminant is biodegraded in the subsurface.

Sorption of organic contaminants onto aquifer solids is frequently described as a partitioning process, where the hydrophobic organic compound partitions into natural organic material associated with the aquifer solids [8]. Sorption can be characterized as either an equilibrium or rate-limited phenomenon. Equilibrium sorption can be modeled as either a linear or non-linear process. Equilibrium sorption may be assumed when the flow of groundwater and other processes affecting contaminant transport are slow compared to the rate of sorption. In this event the sorption of the contaminant can be considered instantaneous. If we assume equilibrium sorption, the relationship between sorbed and aqueous contaminant concentrations may be described by a sorption isotherm.

In many circumstances, however, the assumption of equilibrium sorption is inappropriate. In many laboratory and field studies, contaminant concentration vs time profiles (breakthrough curves) have been observed that exhibited asymmetry. While the assumption of equilibrium sorption results in a prediction of relatively symmetric breakthrough curves, the sharp initial breakthrough and late-tailing curves that are frequently observed are indicative of rate-limited (non-equilibrium) transport [9]. Rate-limited sorption is also a cause of contaminant "rebound" (an increase in contaminant concentration), that is often observed when wells that have been pumping contaminated groundwater are stopped and subsequently restarted [10]. One particular study that was conducted at the Canadian Forces Base in Borden, Ontario, found that the equilibrium assumption was inadequate to describe the subsurface transport of five sorbing compounds. The non-ideal transport was attributed to rate-limited sorption/mass transfer [11–15]. Rate-limited sorption can have a significant effect on the biodegradation of organic contaminants. Since biodegradation occurs in the aqueous phase, the rate at which the contaminant desorbs into the aqueous phase may control the rate of degradation [16, 17]. There are many ways of modeling rate-limited sorption but the two most common are to assume sorption kinetics can be described as either a first-order process or a diffusion-controlled process. Rate-limited sorption should be assumed when sorption is slow relative to the other processes affecting contaminant transport.

2.1.3.1 Equilibrium Sorption

2.1.3.1.1 Linear Model

Most field-scale modeling studies of contaminant plumes make the local equilibrium assumption (LEA) [18, 19]. The LEA is based on the premise that the interactions between the contaminant and the aquifer material are so rapid compared to advective residence times that it can be assumed that the interactions are instantaneous [3]. Linear equilibrium sorption assumes that the binding of contaminants to aquifer solids is instantaneous and that the concentration of sorbed contaminant is directly proportional to the concentration of the dissolved contaminant. This can be modeled by a linear sorption isotherm [2]:

$$S = K_d C \tag{3}$$

where S is the sorbed concentration [M contaminant per M solids], K_d is the sorption partitioning or distribution coefficient [L^3M^{-1}], and C is the dissolved contaminant concentration in the groundwater [ML^{-3}].

2.1.3.1.2 Nonlinear Model

Unlike the linear model, the nonlinear model does not assume that sorbed concentrations are directly proportional to aqueous concentrations. While linear sorption is a good assumption when aqueous contaminant concentrations are relatively low, it may be necessary to use the nonlinear model when concentrations vary over a large range [20]. Nonlinear sorption is often modeled using the Freundlich isotherm, which has been found to empirically describe experimental data [21]:

$$S = K_f C^N \tag{4}$$

where K_f is the Freundlich sorption coefficient $[M^{-N}L^{3N}]$ and N is the Freundlich exponent [-], or the Langmuir isotherm, which assumes sorption to a finite number of sorption sites [21]:

$$S = \frac{Q^0 KC}{KC + 1} \tag{5}$$

where K is the Langmuir sorption coefficient $[L^3M^{-1}]$ and Q^0 is the maximum amount of contaminant that can be sorbed by the solid $[MM^{-1}]$.

2.1.3.2
Rate-Limited Sorption

2.1.3.2.1
Linear Equilibrium

As noted earlier, there is considerable evidence from laboratory and field studies that the LEA is often inadequate to describe reality. When the sorption process occurs at a rate that is similar to the rate of advection, the LEA is not appropriate and rate-limited sorption must be accounted for. Rate limited sorption can be modeled using the following first-order rate expression [21]:

$$\frac{dS}{dt} = k_f C - k_b S \tag{6}$$

where k_f is the first-order sorption rate constant $[L^3M^{-1}T^{-1}]$ and k_b is the first-order desorption rate constant $[T^{-1}]$.

Note that k_f/k_b is equivalent to the sorption distribution coefficient (K_d) in Eq. (3) and at equilibrium (dS/dt=0), Eq. (6) simplifies to the linear equilibrium model, Eq. (3).

2.1.3.2.2
Non-Linear Equilibrium

Rate-limited sorption can also be modeled assuming a kinetic rate expression coupled with a nonlinear equilibrium expression. If we assume a Freundlich isotherm and a first-order rate expression, we can use the following equation to model sorption kinetics [21]:

$$\frac{dS}{dt} = k_f C^N - k_b S \tag{7}$$

where k_f is the first-order sorption rate constant $[M^{-N}L^{3N}T^{-1}]$ and k_b is the first-order desorption rate constant $[T^{-1}]$.

Note that k_f/k_b is equivalent to the Freundlich sorption coefficient (K_f) in Eq. (4) and at equilibrium Eq. (7) simplifies to the Freundlich model, Eq. (4).

If we assume a Langmuir isotherm, with sorption kinetics described by a second-order rate constant (k_f) that depends on both dissolved contaminant concentration and number of available sorption sites, the following expression results [3, 21]:

$$\frac{dS}{dt} = k_f C (Q^0 - S) - k_b S \tag{8}$$

where k_f is the second-order sorption rate constant $[L^3M^{-1}T^{-1}]$ and k_b is the first-order desorption rate constant $[T^{-1}]$.

Here, note that k_f/k_b is equivalent to the Langmuir sorption coefficient (K) in Eq. (5) and at equilibrium Eq. (8) simplifies to Eq. (5).

2.1.3.2.3
Diffusion Model

As noted earlier, diffusion is a process that transports contaminants from areas of high concentration to areas of low concentration. Rate-limited sorption due to diffusion may result from regions of immobile water adjacent to sorption sites so that sorption cannot occur until contaminant diffuses through the immobile water to the site. For example, as contaminated water flows through high hydraulic conductivity regions adjacent to a low conductivity sorptive layer, diffusion moves contaminant from the high hydraulic conductivity region (where concentrations are high) to the low conductivity region (where concentrations are low). Thus, sorption is limited by the diffusive transport process [8]. Diffusion may be modeled using Fick's Second Law of Diffusion [3]. Past research has shown that Fickian diffusion can be modeled using a first-order approximation [22, 23].

2.2
Transformation Processes

We have chosen to follow Watts [24] and discuss chemical and biological transformation processes in the same section. Watts notes that, although this approach is somewhat nontraditional, it is advantageous in that understanding of the abiotic chemical reactions serves as a conceptual basis for understanding the biochemical reactions (which are essentially the same except for the fact that the biochemical reactions are mediated by microorganisms). Where a reaction is predominantly abiotic or biotic, it will be noted in the discussion. In this section, the fundamentals of each chemical or biological reaction will be discussed, and model formulations for the reaction kinetics presented.

2.2.1 Hydrolysis

Hydrolysis, a reaction that typically occurs in the subsurface abiotically, is a substitution reaction in which a compound (RX) reacts with water. In the reaction, the substituent (X) on the compound, which can be a halogen, sulfur, phosphorus or nitrogen, is replaced with a hydroxyl (OH^-) group as shown below [25]:

$$RX + H_2O \rightarrow ROH + HX$$

For example, methyl chloride is hydrolyzed to methanol:

$$CH_3Cl + H_2O \rightarrow CH_3OH + H^+ + Cl^-$$

Hydrolysis results in products, such as alcohols and alkenes, which may be more susceptible to biodegradation – hence hydrolysis may be a significant natural attenuation mechanism.

The more halogenated a compound, the less likely it is to hydrolyze and the slower the reaction kinetics. Hydrolysis of brominated compounds is typically faster than that of chlorinated compounds [26].

Hydrolysis is important in the transformation of 1,1,1-trichloroethane (1,1,1-TCA) to acetic acid [27]. Hydrolysis is also one of the steps in the reaction pathway for degradation of the pesticide dichloro-diphenyl-trichloroethane (DDT) [21].

Hydrolysis reactions can generally be described using first-order kinetics (Eq. 9) where r, the rate of transformation of a contaminant, is proportional to the contaminant concentration (C), and k is a first-order rate constant [26]. Vogel et al. [26] present the half-lives ($t_{1/2}$) for hydrolysis of various halogenated compounds:

$$r = -kC \tag{9}$$

$$t_{1/2} = \frac{\ln (2)}{k} \tag{10}$$

2.2.2 Dehydrohalogenation

Dehydrohalogenation, another abiotic reaction of importance in the subsurface, is an elimination reaction involving halogenated alkanes in which a halogen is removed from one carbon atom, with removal of a hydrogen atom from an adjacent carbon atom and formation of a double bond, as [25]

$$RCXRCHR' \rightarrow RC = CR' + HX$$

Dehydrohalogenation does not include a change of the oxidation state of the organic compound, as there is no net change of charge due to the loss of both a halogen and a hydrogen [26]. Unlike hydrolysis, the likelihood that dehydrohalogenation will occur increases with the number of halogen substituents.

However, as with hydrolysis, bromine substituents are more susceptible than chlorine to undergo dehydrohalogenation [26]. An environmentally important dehydrohalogenation reaction that occurs in the subsurface involves the transformation of 1,1,1-TCA to 1,1-DCE [27].

The reaction rates for dehydrohalogenation have been modeled using first-order-kinetics, and half-lives for transformation of various halogenated compounds are tabulated by Vogel et al. [26].

Butler and Hayes [28] measured pseudo-first-order rate constants for the dehydrohalogenation of several aliphatic compounds in the presence and absence of FeS.

2.2.3 Oxidation Reactions

An oxidation reaction involves the transfer of electrons from a donor, the compound that is oxidized, to an acceptor, the oxidizing agent. Note that the oxidizing agent, which accepts the electrons, is obviously being reduced. However, this redox (reduction-oxidation) reaction is an oxidation reaction in terms of the organic compound of interest, the electron donor, so for our purposes here, the reaction will be referred to as an oxidation reaction. Oxidation reactions can be broadly characterized by the electron acceptor. If the acceptor is oxygen, the reaction is aerobic, while for other acceptors (e.g., nitrate, sulfate, ferric iron, carbon dioxide, etc.) the reaction is anaerobic. Oxidation of organic compounds is typically exothermic, with the greatest energy released when oxygen is the acceptor and the least when carbon dioxide is the acceptor.

2.2.3.1 Aerobic Oxidation

We can categorize microbially-mediated aerobic oxidation reactions as either direct or cometabolic. In a direct reaction, the facilitating microorganism obtains energy and organic carbon from the degradation of the organic compound of interest [25]. A cometabolic reaction requires the presence of a second electron donor (the primary substrate) to serve as the source of energy and carbon. The organic compound of interest is fortuitously oxidized, with no benefit to the facilitating microorganism [29].

2.2.3.1.1 Direct Oxidation

Almost all fuel hydrocarbons (including benzene, toluene, ethylbenzene, and the xylene isomers (BTEX)) can be directly oxidized by microorganisms in the groundwater under aerobic conditions [25]. The following stoichiometry illustrates the direct aerobic oxidation of benzene, resulting in production of microbial cells, carbon dioxide, and water:

$$C_6H_6 + 2.5O_2 + HCO_3 + NH_4 \rightarrow C_5H_7O_2N + 2CO_2 + 2H_2O$$

Typically, direct oxidation of fuel hydrocarbons is modeled as either an instantaneous reaction, or as a reaction that is controlled by first-order:

$$r = -kC \tag{9}$$

or Monod:

$$r = \frac{-k_{max}C}{C + K_M} \tag{11}$$

kinetics [30], where k_{max} is the maximum rate of transformation $[ML^{-3}T^{-1}]$ and K_M is the half-saturation constant $[ML^{-3}]$.

Suarez and Rifai [31] report first-order and Monod rate constants for the aerobic biodegradation of BTEX compounds in laboratory and field studies.

It has also been shown that aromatic hydrocarbon degradation may be self-inhibited. That is, the rate of degradation of the hydrocarbon is slowed by high concentrations of the hydrocarbon itself [29, 32]. Self-inhibition may be modeled by incorporating Haldane (or Andrews) kinetics into the Monod expression:

$$r = \frac{-k_{max}C}{C + K_M + \dfrac{C^2}{K_I}} \tag{12}$$

where K_I is the Haldane inhibition concentration of the self-inhibitory compound $[ML^{-3}]$.

Direct oxidation of the lesser chlorinated ethenes, ethanes, polychlorinated benzenes, and chlorobenzene has been reported. Wiedemeier et al. [25] summarize a number of studies that report direct aerobic oxidation of vinyl chloride (VC), 1,2-dichloroethane, the three dichlorobenzene isomers, 1,2,4-trichlorobenzene, and 1,2,4,5-tetrachlorobenzene. Bradley [33] reports that DCE has served as a primary substrate for energy production with oxygen as the electron acceptor, though use of DCE as a sole carbon source has not been demonstrated. Rittmann and McCarty [29] also report that the two least chlorinated methanes (dichloromethane and chloromethane) as well as chloroethane can be directly oxidized under aerobic conditions. Direct oxidation of the chlorinated compounds is typically modeled using either first-order or Monod kinetics [29, 31].

2.2.3.1.2
Cometabolism

With the recent finding that PCE can be aerobically cometabolized [34], it has now been shown that except for carbon tetrachloride, all of the major chlorinated ethenes, ethanes, and methanes found in groundwater may be biodegraded by this process [29]. As noted earlier, aerobic cometabolism involves direct oxidation of an electron donor (the primary substrate) with oxygen as the electron acceptor. This direct oxidation provides energy and organic carbon to the facilitating microorganisms, which produce an enzyme to catalyze the reaction. Because the enzyme produced to catalyze the oxidation of the primary substrate is non-specific, the contaminant of interest is fortuitously oxidized as well. Aerobic cometabolism of trichloroethylene (TCE), the most frequently detected groundwater contaminant at hazardous waste sites in the U.S. [35], has been the subject of considerable research. Laboratory studies of TCE cometabolism have shown that methane, phenol, toluene, propane, ethylene, and other compounds can all serve as primary substrates to induce native bacteria to produce oxygenase enzymes that fortuitously oxidize TCE to TCE epoxide [36, 37]. The cometabolic degradation reactions for TCE, with methane as the primary substrate and methane monooxygenase (MMO) as the nonspecific enzyme that catalyzes the TCE epoxidation, are shown in Fig. 1.

Aerobic cometabolism also may play a role in the oxidation of 1,1,1-TCA, where a hydroxylation reaction is the initial step [27]:

$$CH_3CCl_3 + H_2O \rightarrow HOCH_2CCl_3 + 2H^+ + 2e^-$$

Modeling cometabolic degradation involves a set of equations that describes microbial growth due to degradation of the primary substrate, and degradation of the contaminant of interest due to cometabolism. Cometabolic degradation has been modeled using a dual-Monod kinetic expression where degradation of the primary substrate (the electron donor) and target contaminant are dependent on Monod kinetic terms for both the electron donor and acceptor [38]. In addition, it is necessary to account for competitive inhibition in the model, since the primary substrate and target contaminant compete for the enzyme that catalyzes both the direct oxidation and cometabolic oxidation reactions. It has also been shown that the cometabolic epoxidation reaction is, in fact, harmful to the bacteria that facilitate it. This transformation product toxicity phenomenon is modeled using a transformation capacity (T_c) parameter, defined as the quantity of the target contaminant that a given mass of bacteria

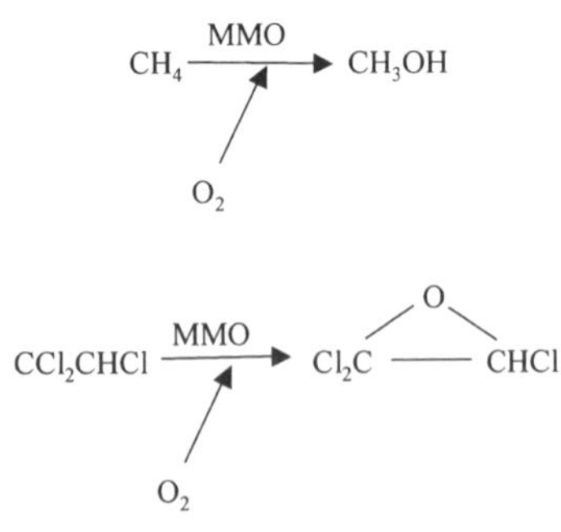

Fig. 1 Aerobic cometabolism of TCE with methane as the primary substrate

can transform before being destroyed by the transformation [29]. Finally, it has been seen that when bacteria are not growing, the bacteria "deactivate", that is, the fraction of biomass that is active toward degradation of the target contaminant decreases [38]. The equations below model growth of bacteria (X) resulting from aerobic oxidation of a primary substrate (C_D), and cometabolism of a target contaminant (C), while accounting for competitive inhibition, transformation product toxicity, and biomass deactivation [38, 39]:

$$\frac{dX}{dt} = k_{\max} YX \left(\frac{C_D}{K_{MD}\left(1 + \frac{C}{K_M}\right) + C_D} \right) \left(\frac{C_A}{K_{MA} + C_A} \right) - bX \left(\frac{C_A}{K_{MA} + C_A} \right) + \frac{r}{T_C} \quad (13)$$

$$\frac{dC_D}{dt} = -k_{\max} X \left(\frac{C_D}{K_{MD}\left(1 + \frac{C}{K_M}\right) + C_D} \right) \left(\frac{C_A}{K_{MA} + C_A} \right) \quad (14)$$

$$r = -k_T F_A X \left(\frac{C}{K_M \left(1 + \frac{C_D}{K_{MD}}\right) + C} \right) \left(\frac{C_A}{K_{MA} + C_A} \right) \quad (15)$$

$$\frac{dF_A}{dt} = -b_d F_A \; if \; \frac{dX}{dt} < 0, otherwise \; F_A = 1$$

where Y is the yield coefficient [MM^{-1}], k_{max} is the maximum rate of transformation of the electron donor [T^{-1}], k_T is the maximum rate of transformation of the target contaminant [T^{-1}], K_{MD} is the half-saturation constant for the donor [ML^{-3}], K_M is the half-saturation constant for the target contaminant [ML^{-3}], K_{MA} is the half-saturation constant for the acceptor [ML^{-3}], b is the biomass decay coefficient [T^{-1}], T_C is the transformation capacity [MM^{-1}], F_A is the fraction of biomass that is active toward TCE degradation, and b_d is the rate constant for deactivation of biomass [T^{-1}].

In the above dual-Monod equations, we note that competitive inhibition between the primary substrate and target contaminant is modeled by the C/K_M and C_D/K_{MD} terms in the denominator on the right hand side of Eqs. (13)–(15), transformation product toxicity is modeled by the last term in Eq. (13), and deactivation is modeled by the F_A parameter in Eq. (15).

2.2.3.2
Anaerobic Oxidation

It has been demonstrated that BTEX compounds can readily be oxidized using nitrate as an electron acceptor [29, 40–42]. With only a few exceptions (for instance, ethylbenzene has not been demonstrated to degrade under sulfate-reducing conditions), BTEX oxidation using sulfate, carbon dioxide, ferric iron, and Mn(IV) as electron acceptors has also been demonstrated [29, 40]. As would be expected based on energy considerations, anaerobic BTEX oxidation is slower than aerobic oxidation. First-order BTEX oxidation rate constants under anaerobic conditions are reported in Suarez and Rifai [31].

Anaerobic oxidation has also been demonstrated as a pathway for chlorinated ethene degradation. Microorganisms have been shown to oxidize VC to CO_2 under ferric iron-reducing conditions [43] and DCE has been shown to oxidize directly to CO_2 under Mn(IV)-reducing conditions [44]. In addition, Bradley [33] reports that VC can be oxidized under methanogenic conditions. Suarez and Rifai [31] report first-order rate constants for the anaerobic oxidation of VC.

2.2.4 Reduction Reactions

2.2.4.1 Hydrogenolysis

Hydrogenolysis is the replacement of a halogen substituent on an organic compound by a hydrogen as

$$RX + H^+ + 2e^- \rightarrow RH + X^-$$

Halogenated ethenes hydrogenolysize more readily than halogenated ethanes. In Fig. 2 the biotic transformation of PCE to ethene by sequential hydrogenolysis is depicted. Biologically mediated hydrogenolysis can be either a cometabolic or direct process. As a cometabolic process, the terminal electron acceptor from which the microorganism derives energy is not the halogenated compound and the halogenated compound is reduced cometabolically. As a direct process,

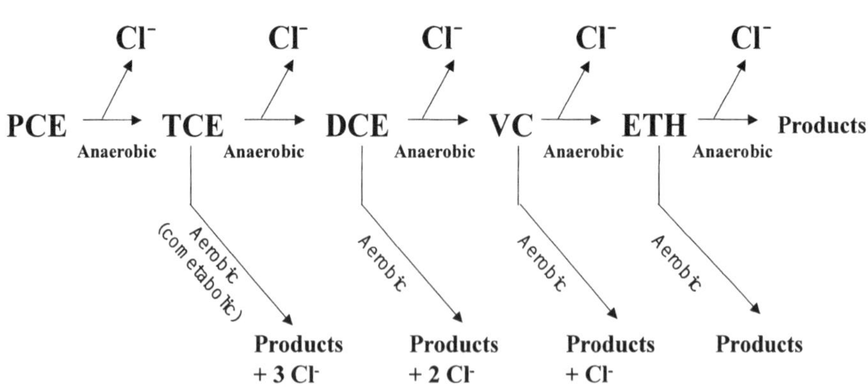

Fig. 2 Pathways modeled for aerobic and anaerobic degradation of PCE and TCE at Dover Air Force Base (after [19])

termed halorespiration, the halogenated compound serves as the terminal electron acceptor and the halorespiring microorganism gains energy directly from reducing the compound [29, 33]. The more reducing the environment, the more readily will a compound hydrogenolysize. For example, 1,1,1-TCA has been shown to undergo hydrogenolysis under anaerobic conditions, and most rapidly under methanogenic conditions [27]:

$$CH_3CCl_3 + H^+ + 2e^- \rightarrow CH_3CHCl_2 + Cl^-$$

Butler and Hayes [28] modeled the rate of abiotic hydrogenolysis of hexachloroethane (HCA), 1,1,1-TCA, carbon tetrachloride (CT), and tribromomethane (TBM) with FeS as a reducing agent using pseudo-first-order kinetics.

The reductive dechlorination of alachlor, metolachlor, and DDT, three important pesticides, was observed to occur by hydrogenolysis [21, 45]. First-order degradation rate constants were reported to be 0.12 and 0.10 h^{-1} for 10 mg/l solutions of alachlor and metochlor, respectively.

Alachlor is also known to be biotransformed under anaerobic conditions, with acetyl alachlor the transformation product of hydrogenolysis [46]. Transformation kinetics were modeled as a second-order reaction:

$$r = -kC\,[\text{Biomass}] \tag{16}$$

where the second-order biotransformation coefficients for alachlor (k) were $7.6\times10^{-5}(\pm4.0\times10^{-5})$ under denitrifying conditions, $2.9\times10^{-3}(\pm1.6\times10^{-3})$ under methanogenic conditions, and $1.5\times10^{-2}(\pm1.4\times10^{-2})$ l mg VSS^{-1} day^{-1} under sulfate-reducing conditions.

2.2.4.2 Dihaloelimination

Dihaloelimination is the removal of two halogens and the formation of a double carbon-carbon bond as [25]

$$RCXCXR' + 2e^- \rightarrow RC = CR' + 2X^-$$

For example, 1,1,1,2-tetrachloroethane (1,1,1,2-TeCA) is transformed into 1,1-dichloroethylene by dihaloelimination [28]. Butler and Hayes [28] tabulate pseudo-first-order reaction rates for dihaloelimination of TeCA, HCA, pentachlorethane (PCA), and 1,1,2-TCA.

2.2.4.3 Reduction of Explosives

Explosives, such as 2,4,6-trinitrotoluene (TNT), hexahydro-1,3,5,-trinitro-1,3,5-triazine (RDX), and octahydro-1,3,5,7-tetranitro-1,3,5,7-tetrazocine (HMX), present a serious subsurface environmental hazard throughout the world. A number of studies on the biological reduction of these compounds have been reported. Most of these studies used first-order-kinetics to model reduction kinetics [47–51].

Hofstetter et al. [52] also demonstrated polynitroaromatics, such as TNT, can be completely reduced to the aromatic polyamines by ferrous ion present at the surface of ferric oxides, while they can be less efficiently transformed by the hydroquinones of natural organic matter in the presence of H_2S. They modeled this transformation as a zero-order reaction, where

$$r = -k \tag{17}$$

3 Modeling

Models may be used to gain better understanding and insight into the various processes at work in the subsurface, as well as to quantitatively answer the important questions: how much contaminant will be transported from a source to a receptor, and in what timeframe will the transport occur? As seen in the previous sections, there are many interrelated processes occurring in the subsurface. A model helps to demonstrate how these processes interact, as well as to show which processes are most important in determining contaminant fate and transport. Mathematical models use equations to simulate groundwater flow and contaminant transport in the subsurface. The basic equation that is used to model contaminant fate and transport in groundwater is the advective/dispersive reactive transport equation, which is discussed in the next section. This equation, along with associated boundary and initial conditions, may be solved analytically or numerically. As discussed below, an exact analytical solution may only be obtained when the equation and initial/boundary conditions have been greatly simplified. For instance, an analytical solution could not be obtained if the model equation is nonlinear (such as when sorption is described using the Freundlich or Langmuir isotherms defined in Eqs. (4) and (5). On the other hand, as discussed further below, numerical methods can be used to solve nonlinear equations with complex boundary and initial conditions. However, these methods only provide an approximate solution to the governing equations. Due to the complexity of the subsurface environment, most models that are used to simulate contaminant transport are numerical. There are a number of numerical model codes that have been developed to simulate fate and transport of chemicals in the subsurface. A few of the most commonly used model packages will be discussed later. Finally, the section closes with examples of how models may be applied to manage contaminated sites.

3.1
General Advective/Dispersive Reactive Transport Equation

As contaminant transport occurs over times much greater than the times over which groundwater flow fluctuates, steady flow is frequently assumed. For steady groundwater flow in three dimensions, the following vector equation, developed based on mass conservation principles, is typically used to model advective/dispersive transport of a dissolved reactive contaminant (after [53]):

$$\frac{\partial C}{\partial t} = \nabla \bullet (\tilde{\tilde{D}} \bullet \nabla C) - \tilde{v} \bullet \nabla C - \frac{\rho_b}{n}\frac{\partial S}{\partial t} - r \tag{18}$$

where $\tilde{\tilde{D}}$ is the dispersion coefficient matrix $[L^2T^{-1}]$, $\tilde{v}$ is the groundwater average linear velocity vector $[LT^{-1}]$, ρ_b is the bulk density of aquifer solids $[ML^{-3}]$, and r is the sink due to chemical/biological transformation of contaminant $[ML^{-3}T^{-1}]$.

The first term on the right-hand-side of the equation accounts for contaminant dispersion in the x-, y-, and z-directions, while the second term accounts for contaminant advection. The third term on the right-hand-side of the equation is a sink term to account for sorption of dissolved contaminant to aquifer solids, and the fourth term is a sink term to account for loss of dissolved contaminant mass due to chemical and biological reactions. Note that one of the implicit assumptions in Eq. (18) is that the chemical/biological reaction sink applies only to dissolved phase contaminant. This is in accord with the frequently made observation that sorbed contaminant is not bioavailable [5–7].

Recall from Eq. (2) that the dispersion matrix depends on velocity. Thus, for the first and second terms on the right-hand-side of Eq. (18), the groundwater average linear velocity vector (which was assumed steady in time) must be determined. This is accomplished in two-steps. In the first step, the distribution of hydraulic heads must be determined in order to calculate the hydraulic gradient, for use in Eq. (1). For steady flow, the head field must satisfy Laplace's equation; that is

$$\nabla^2 h = 0 \tag{19}$$

For specified hydraulic head boundary conditions, Equation (19) can be solved for h (x,y,z) at all points in space. In the second step, knowing h (x, y, z), we may apply Darcy's Law, Eq. (1), to calculate $\tilde{v}$ (x, y, z) for use in Eq. (18).

Equations (6)–(8) may be substituted into Eq. (18) to describe first-order rate-limited sorption for linear, Freundlich, and Langmuir isotherms, respectively. If equilibrium sorption can be assumed, Eqs. (3)–(5) may be used to define S in Eq. (18) in terms of C for linear, Freundlich, and Langmuir isotherms, respectively [21].

Depending on the relevant chemical/biological process(es) assumed to be affecting contaminant fate, the last term on the right-hand-side of Eq. (18) can be modeled using the applicable equation(s) presented above.

3.2 Model Solutions to the Transport Equation

Equation (18) is a partial differential equation which, for given initial and boundary conditions, can be solved for contaminant concentration as a function of space and time. Both analytical and numerical solution methods will be discussed below.

3.2.1 Analytical

For a linear partial differential equation with relatively simple initial and boundary conditions, and uniform parameter values, an analytical solution may be obtained. Although simplified, analytical solutions can be put to a number of important uses. Analytical models may be used to [9]:

1. Estimate contaminant exposure concentrations over space and time for a given source
2. Fit concentration data to obtain parameter estimates for use in more complex models
3. Determine allowable source concentrations based upon permissible downgradient contaminant exposure concentrations
4. Calculate the transport component of a risk assessment model
5. Evaluate the performance of numerical models

In Table 1, a number of 3-D analytical models are listed. Depending on the situation being modeled, and the modeling objective, it may be more appro-

Table 1 3-D analytical models

Initial/boundary conditions	Processes modeled [a]	Reference
Instantaneous point source	Advection, dispersion, first-order decay	[54]
Instantaneous and continuous point sources, instantaneous parallelepiped source	Advection, dispersion	[55]
Continuous rectangular source	Advection, dispersion, first-order decay	[56]
Horizontal planar, time varying source	Advection, dispersion, first-order decay	[57]
Instantaneous point source and instantaneous sources of various simple geometries	Advection, dispersion, rate-limited sorption	[58]
Instantaneous point and parallelepiped sources	Advection, dispersion, rate-limited sorption	[12]

[a] All solutions also applicable for linear, equilibrium sorption.

priate to use 1- or 2-D models in a particular instance. 3-D models are tabulated here, as derivation of the 1- and 2-D analog of a 3-D model is straightforward.

3.2.2 Numerical

Numerical methods allow one to obtain a solution to the groundwater contaminant transport equation even when the equation is nonlinear (for example, when sorption is described by either a Freundlich or Langmuir isotherm). These numerical methods also permit solution of the governing partial differential equation when parameter values and initial and boundary conditions vary in space and time.

Obtaining numerical solutions to the model partial differential equation first involves approximating the governing equation, along with the equation's initial and boundary conditions, by some technique to obtain a set of algebraic equations and then numerically finding a solution to the algebraic equations. The most common techniques used in groundwater contaminant transport models to approximate the governing equations are: (1) finite difference methods, (2) finite element methods, and (3) the method of characteristics. While we will not discuss details of the various techniques, and their advantages and disadvantages, the reader is referred to the literature for discussions on numerical solutions to the groundwater transport equation (e.g., [59–61]).

3.2.3 Model Codes

There are a number of model codes that have been developed that allow users to conveniently apply analytical and numerical solutions of the governing transport equations to various contamination scenarios. The following discussion will look at the capabilities and limitations of some of these codes. We will restrict our discussion to the more popular numerical codes that can simulate 2- or 3-D contaminant transport, as analytical models and 1-D solutions are usually too simplified for application to real contamination scenarios.

3.2.3.1 MODFLOW/MT3D

MODFLOW is a very popular 3-D code in the public domain, developed by the U.S. Geological Survey (USGS), that is used to model groundwater flow [62]. MODFLOW uses finite differences to solve the steady-state groundwater flow equation (Eq. 19) for specified boundary conditions. The output of MODFLOW is the spatial distribution of hydraulic heads, which may then be used in Darcy's equation to determine groundwater velocity at all points in space. This velocity output file may be used as an input file by several contaminant transport codes [60]. Note that MODFLOW also can solve the transient groundwater flow equation for specified initial and boundary conditions. As mentioned earlier, however, steady-state flow can frequently be assumed for contaminant transport problems.

The first transport code to make use of MODFLOW's head and flow output files was MT3D [63]. MT3D allows the user to specify which solution technique is applied to solve the contaminant transport equation. Depending upon the problem conditions, the user can choose either the finite difference method or from among various adaptations of the method of characteristics to approximate the governing transport equation. MT3D simulates advection, dispersion, first-order degradation, and equilibrium sorption (described using a linear, Freundlich, or Langmuir isotherm).

3.2.3.2
Method of Characteristics (MOC) and MOC3D

The original MOC code, which as the name suggests relies on the method of characteristics to approximate the contaminant transport governing equation, was also developed by the USGS [64]. The MOC code solves both the groundwater flow equation and the advective/dispersive transport equation in 2-D. In 1996, a 3-D version of the MOC, called MOC3D, was published [65]. Unlike MOC, which solves the flow equation internally, MOC3D is integrated with MODFLOW, so that MODFLOW head and groundwater flow output files are used as input for MOC3D. MOC3D simulates advection, dispersion, linear equilibrium sorption, diffusion into zones of immobile water (using a first-order approximation), and zeroth and first-order growth and/or degradation [66].

3.2.3.3
BIOPLUME III

BIOPLUME III is a public domain transport code that is based on the MOC (and, therefore, is 2-D). The code was developed to simulate the natural attenuation of a hydrocarbon contaminant under both aerobic and anaerobic conditions. Hydrocarbon degradation is assumed due to biologically mediated redox reactions, with the hydrocarbon as the electron donor, and oxygen, nitrate, ferric iron, sulfate, and carbon dioxide, sequentially, as the electron acceptors. Biodegradation kinetics can be modeled as either a first-order, instantaneous, or Monod process. Like the MOC upon which it is based, BIOPLUME III also models advection, dispersion, and linear equilibrium sorption [67].

3.2.3.4
RT3D

RT3D is an extension of MT3D, developed by researchers at Battelle Pacific Northwest Laboratories to simulate natural attenuation of chlorinated contaminants, that is capable of modeling multi-species reactive transport [19, 68].

As an extension of MT3D, it is a 3-D advective/dispersive transport model that uses MODFLOW to solve the flow equation. As a multi-species model, RT3D can model sequential decay reactions, so that transport of parent and daughter compounds is simulated. RT3D is extremely flexible, allowing the user to specify almost any type of degradation and sorption kinetic submodel for incorporation as a reaction package into the code [60]. RT3D also has preprogrammed reaction packages for many of the reactions discussed above (e.g., rate-limited sorption, Monod and dual-Monod kinetics, etc.). The flexibility of RT3D comes at a price, however, as the user must be fairly conversant with the code and modeling techniques to incorporate reaction packages [69].

3.2.3.5 BioRedox

Like RT3D, BioRedox is a 3-D model that is capable of modeling multi-species reactive transport [70]. The public domain model can simulate coupled oxidation-reduction reactions between multiple electron acceptors and donors. Except for rate-limited sorption, it is capable of simulating all the reactions simulated by RT3D, and is more user-friendly, in that no modifications to source code are required to incorporate reaction packages [70].

3.2.3.6 BIO3D

BIO3D is a 3-D numerical model that, like RT3D and BioRedox, also simulates multi-species reactive transport. Unlike BioRedox, however, BIO3D only accommodates a single electron acceptor [71]. Sorption in BIO3D is modeled as a simple linear equilibrium process. Growth of attached microorganisms is assumed to be governed by dual-Monod kinetics, modified by a Haldane inhibition term.

3.3 Case Studies of Model Applications

Probably the best examples of model application to simulate organic chemical fate and transport in the subsurface come from studies that have been conducted in recent years where models have been used to simulate natural attenuation of chlorinated solvents and petroleum hydrocarbons in groundwater. Natural attenuation, which relies on naturally occurring physical, chemical, and biological processes to achieve site-specific remedial objectives, has shown great promise as a groundwater contamination remediation strategy, particularly with regard to remediation of petroleum hydrocarbon contamination [72]. As incidents of petroleum hydrocarbon and chlorinated solvent groundwater contamination are the most prevalent, considerable effort has been spent investigating whether natural attenuation can be applied to these two classes of chemicals. A significant component of these investigations has been modeling the natural attenuation processes at various sites, to see whether natural attenuation is a viable remediation strategy at those sites. In the case studies that follow, various models are applied at sites contaminated with petroleum hydrocarbons and chlorinated solvents, to simulate natural attenuation processes. The goal of the following discussion is to ascertain whether our current modeling capabilities and understanding of processes are adequate to approximate the impact of natural subsurface attenuation processes on contaminant fate and transport.

3.3.1 Modeling Petroleum Hydrocarbon Fate and Transport During a Field Experiment at Canadian Forces Base (CFB) Borden

Schirmer et al. [71] applied the BIO3D model to simulate fate and transport of petroleum hydrocarbons (benzene, toluene, ethylbenzene, and the xylene (BTEX) isomers) under aerobic conditions during a field experiment conducted at CFB Borden. The model simulated BTEX fate and transport using the advection/dispersion equation, with sink terms accounting for linear equilibrium sorption and biodegradation. Biodegradation of the BTEX compounds was governed by dual-Monod kinetics under oxygen limiting conditions, with Haldane inhibition, as indicated below:

$$r_i = k_{\max,i} X \frac{C_i}{C_i + K_{Si} + \dfrac{C_i^2}{K_{Ii}}} \frac{C_{Oxygen}}{C_{Oxygen} + K_{Soxygen}} \qquad (20)$$

where r_i is the biodegradation rate of the i-th BTEX compound [$ML^{-3}T^{-1}$], C_i is the dissolved concentration of the ith BTEX compound [ML^{-3}], C_{oxygen} is the dissolved concentration of oxygen [ML^{-3}], X is the attached microbial concentration [ML^{-3}], K_{Si} is the half-utilization constant for the ith BTEX compound [ML^{-3}], $K_{Soxygen}$ is the half-utilization constant for oxygen [ML^{-3}], k_{maxi} is the maximum utilization rate of the ith BTEX compound [T^{-1}], and K_{Ii} is the Haldane inhibition concentration of the ith BTEX compound [ML^{-3}].

Interaction between the different BTEX compounds is accounted for through the microbial growth term, as follows:

$$\frac{dX}{dt} = \sum_{i=1}^{N} k_{\max,i} X Y_i \left(1 - \frac{X}{X_{\max}}\right) \left(\frac{C_i}{C_i + K_{Si} + \dfrac{C_i^2}{K_{Ii}}} \right) \left(\frac{C_{Oxygen}}{C_{Oxygen} + K_{Soxygen}} \right) - bX \qquad (21)$$

where Y_i is the microbial yield per unit of the ith BTEX compound consumed [MM^{-1}], N is the number of compounds, X_{max} is the maximum microbial concentration [ML^{-3}], and b is the first-order decay rate of the microbial population [T^{-1}].

Using independently determined parameters from the laboratory and prior field studies at the site, the study authors were able to apply the uncalibrated model to match experimental results reasonably well.

3.3.2 Modeling Petroleum Hydrocarbon Fate and Transport at a Contaminated Site on Patrick Air Force Base

Rifai et al. [30] applied the 2-D BIOPLUME III to simulate fate and transport of BTEX at a contaminated groundwater site at Patrick Air Force Base. Whereas in the study described above, the BTEX compounds were modeled as multiple electron donors with a single electron acceptor (oxygen), in the Rifai et al. [30] study, BTEX is modeled as a single electron donor, with multiple electron acceptors (oxygen, ferric iron, and carbon dioxide). Although BIOPLUME III has the capability of simulating degradation kinetics as either a first-order or Monod process, for the Patrick study, biodegradation was modeled as an instantaneous reaction between electron donor and acceptors (implicitly assuming that the biodegradation reaction rate is much greater than the rates of other physical and chemical processes impacting BTEX fate and transport). Sorption was modeled as a linear equilibrium process.

Using hydraulic conductivity and BTEX source injection rate and concentration as fitting parameters; the model was calibrated to field data and the authors found "reasonable agreement" between the calibrated model simulation and field data. The model was then applied in a predictive mode, and simulations showed the area of contamination expanding slightly and then stabilizing, indicating that natural attenuation may be a viable strategy at the site.

3.3.3 Modeling Chlorinated Ethene Fate and Transport at a Contaminated Site on Dover Air Force Base

Clement et al. [19] used the 3-D RT3D code to simulate fate and transport of chlorinated ethene compounds at a contaminated groundwater site at Dover Air Force Base. The two primary contaminants that were spilled at the site were PCE and TCE. However, presumably due to reductive dehalogenation, DCE, vinyl chloride (VC), ethene (ETH), and chloride (Cl^-) are also present as groundwater contaminants.

In the RT3D simulation, advective/dispersive transport of each contaminant is assumed. Sorption is modeled as a linear equilibrium process and biodegradation is modeled as a first-order process. Due to the assumed degradation reaction pathways (Fig. 2) transport of the different compounds is coupled. In the study, four reaction zones were delineated, based on observed geochemistry data. Each zone (two anaerobic zones, one transition zone, and one aerobic zone) has a different value for the biodegradation first-order rate constant for each contaminant. For example, since PCE is assumed to degrade only under anaerobic conditions, the first-order rate constant for PCE reductive dehalogenation to TCE in the aerobic zone is set at zero. For TCE, which can be both reductively dehalogenated under anaerobic conditions and aerobically cometabolized, the first-order rate constant for reductive dehalogenation is higher in the zones that are more reducing and the rate constant for aerobic cometabolism increases as oxidizing conditions increase.

Using contaminant sources and reaction rate constants as fitting parameters, while constraining reaction rates to literature values, the authors calibrated the model to the data and obtained "good agreement." In addition, for the most part, the fitted rate constants were close to field and/or laboratory measured values. The authors were careful to note, however, that the model was not used in a predictive mode.

3.3.4
Modeling Chlorinated Ethene Fate and Transport at a Contaminated Site on F.E. Warren Air Force Base

Young et al. [73] used a non-commercial research code [74] to simulate degradation of TCE at a contaminated groundwater site at F.E. Warren Air Force Base. The code has the capability of simulating 3-D advection and dispersion, linear and non-linear equilibrium sorption, rate-limited sorption, and biodegradation assuming first-order, Monod, or Dual Monod kinetics. For the F.E. Warren study, sorption was simulated as a linear equilibrium process and biodegradation was assumed to involve sequential reductive dehalogenation of TCE to DCE to VC to ETH. The biodegradation reactions were modeled using first-order kinetics. The authors used site data from 1993 to calibrate the model, using the biodegradation reaction first-order rate constants as fitting parameters. The model was then run in a predictive mode, to predict concentrations of TCE, DCE, and VC in 1999. A comparison of predicted and measured 1999 contaminant concentrations is depicted in Figs. 3, 4, 5. The authors also present goodness-of-fit statistics, in an attempt to quantify how well predicted and measured contaminant concentrations agree. Based on these statistics, the authors characterized the model predictions as good for TCE, fair for DCE, and due to a lack of measured data, no conclusions were drawn regarding VC predictions.

3.3.5
Modeling Petroleum Hydrocarbon and Chlorinated Ethene Fate and Transport at a Contaminated Site on Plattsburgh Air Force Base

Carey et al. [70] used the BioRedox code to simulate the fate and transport of BTEX and chlorinated ethenes at a contaminated groundwater site at Plattsburgh Air Force Base. Transport of the compounds was modeled using the 3-D advection/dispersion equation, and sorption was assumed to be negligible. While BioRedox is capable of simulating oxidation of multiple electron donors,

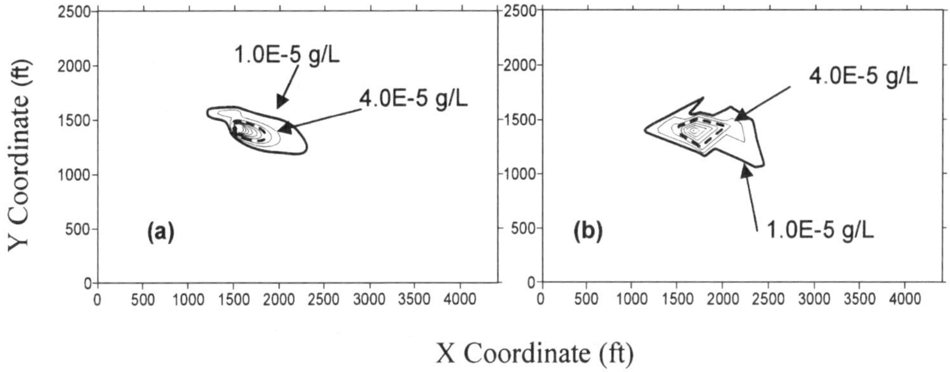

Fig. 3a, b TCE concentrations at F.E. Warren AFB for 1999: **a** simulated; **b** observed

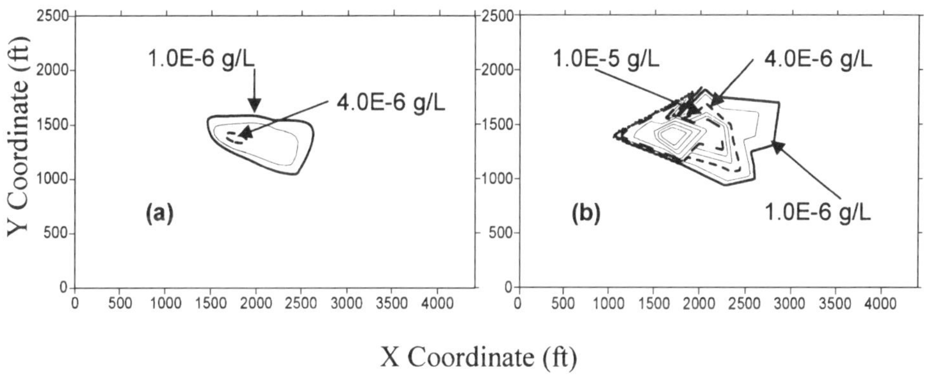

Fig. 4a, b DCE concentrations at F.E. Warren AFB for 1999: **a** simulated; **b** observed

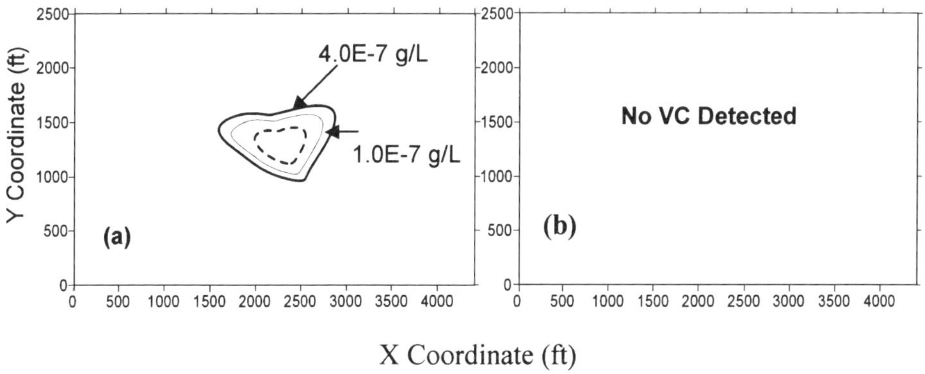

Fig. 5a, b VC concentrations at F.E. Warren AFB for 1999: **a** simulated; **b** observed

BTEX was represented as a single electron donor in this study. Electron acceptors assumed for the study were oxygen, nitrate, ferric iron, sulfate, and carbon dioxide. BTEX oxidation was assumed to be instantaneous in the aerobic region, and to be governed by first-order kinetics in the anaerobic regions. TCE, DCE, and VC were all assumed to be cometabolically biodegraded in the aerobic region, so long as methane, which would be produced under methanogenic conditions and whose transport was simulated, was present as a primary substrate. The cometabolic biodegradation reaction was modeled using first-order kinetics. Under methanogenic conditions, reductive dehalogenation of TCE to DCE to VC to ETH was assumed to occur, modeled as a series of first-order reactions. While TCE was assumed to also be reductively dehalogenated under sulfate- and ferric iron-reducing conditions, DCE and VC were assumed to not be reductively dehalogenated under those conditions. VC could be oxidized, however, under both aerobic and ferric iron reducing conditions.

Using literature values for the various biodegradation first-order rate constants, concentrations over space and time for all the contaminants (as well as methane, chloride, and ferrous iron) were simulated. BioRedox includes a visualization module that simplifies comparison between model predictions and field measurements for the numerous compounds of interest. While the authors give an example of predicted versus measured results at several locations on the site, no attempt is made to quantify the quality of the overall prediction.

4 Discussion

As the above case studies demonstrate, our capability to model the behavior of organic chemicals in groundwater is substantial, yet today we still remain far from being able to confidently predict contaminant concentration response at places and times very far removed from the source of contamination. Additional study is needed so that we may be able to (1) extend what we know in the laboratory about the basic physicochemical and transformation processes to the field and (2) understand more about the basic processes.

4.1 Extending Laboratory Knowledge to the Field

Even though a process may be well understood, modeling how that process interacts with a contaminant in a field setting involves considerable difficulty. Physical, chemical, and biological conditions in the subsurface vary both spatially and temporally, and these heterogeneities may need to be accounted for to simulate contaminant fate and transport. Also, parameter values to be used in the model must be determined, and oftentimes, the values that may be obtained in the laboratory or from the literature are not representative of field conditions. Finally, as discussed earlier, certain processes like dispersion have been demonstrated to be scale-dependent, so that the value of a parameter like the dispersivity is a function of the scale of the problem under consideration [2].

4.2 Understanding Basic Processes

While we have adequate ability to model individual processes (e.g., sorption, advection, biodegradation), we encounter difficulties when we attempt to model interactions between processes. For example, microbial growth due to biodegradation of a subsurface contaminant may result in reduction of hydraulic conductivity, leading to a change in groundwater flow. This interaction between biological growth and flow is an area of current research (e.g., [75]). The interaction between sorption and biodegradation phenomena is also not well understood. As mentioned in the discussion below Eq. (18), it is typically assumed that the rate of contaminant transformation (r) is dependent on dissolved phase contaminant concentration, as sorbed contaminant is assumed not to be bioavailable. However, recent experimental work has indicated that sorbed contaminant may be susceptible to biodegradation [76]. It has also been hypothesized that sorption/biodegradation may be coupled, with the presence of microbial cells serving to either enhance or suppress sorption/desorption processes [77].

As may be apparent from many of the examples presented above, models of processes that affect the fate and transport in groundwater of petroleum hydrocarbons and chlorinated aliphatic hydrocarbons are much more prevalent than models of other contaminants. Much research has yet to be accomplished to improve our understanding and ability to model transformation processes, particularly for contaminants like polynuclear aromatic hydrocarbons, nitroaromatics, and pesticides.

Acknowledgments Mark Goltz did this work while on sabbatical at the National Subsurface Environmental Research Laboratory (NSERL) at Ewha Womans University. The sabbatical was supported by the Air Force Institute of Technology, the Air Force Office of Scientific Research Windows on Asia program, and the NSERL. Yu-Jin Chung, In-Suk Mook, and Ji-Hyun Kim contributed to this work as part of their senior research project. The views expressed in this article are those of the authors and do not necessarily reflect the official policy or position of the Air Force, The Department of Defense, or the U.S. Government.

References

1. Chrysikopoulos CV (2005) Nonaqueous Phase Liquid Pool Dissolution in Subsurface Formations In: Kassim TA (ed) Environmental Impact Assessment of Recycled Wastes on Surface and Groundwaters: (Volume 3) Engineering Modeling and Substainability. Handbook of Environmental Chemistry, Water Pollution Series, Vol 5/Part F. Springer, Berlin Heidelberg New York: 97–132

2. Domenico PA, Schwartz FW (1998) Physical and chemical hydrogeology, 2nd edn. Wiley, New York
3. Clark MM (1996) Transport modeling for environmental engineers and scientists. Wiley, New York
4. Gelhar LW, Welty C, Rehfeldt KR (1992) A critical review of data on field-scale dispersion in aquifers. Water Resour Res 28:1955–1974
5. Ramaswami A, Luthy RG (1997) Measuring and modeling physicochemical limitations to bioavailability and biodegradation In: Hurst CJ, Knudsen GR, McInerney MJ, Stetzenbach LD, Walter MV (eds) Manual of environmental microbiology. ASM Press, Washington DC, pp 721–729
6. Alexander M (2000) Aging, bioavailability, and overestimation of risk from environmental pollutants. Environ Sci Technol 34:4259–4265
7. Zhao X, Voice TC (2000) Assessment of bioavailability using a multicolumn system. Environ Sci Technol 34:1506–1512
8. Werth CJ (2005) Equilibrium Partitioning and Mass Transfer of Organic Chemicals Leached from Recycled Hazardous Waste Materials. In: Kassim TA (ed) Environmental Impact Assessment of Recycled Wastes on Surface and Groundwaters: (Volume 3) Engineering Modeling and Substainability. Handbook of Environmental Chemistry, Water Pollution Series, Vol 5/Part F. Springer, Berlin Heidelberg New York: 1–32
9. Charbeneau RJ (2000) Groundwater hydraulics and pollutant transport. Prentice Hall, Upper Saddle River, New Jersey
10. Harvey CF, Haggerty R, Gorelick SM (1994) Aquifer remediation: a method for estimating mass transfer rate coefficients and an evaluation of pulsed pumping. Water Resour Res 30:1979–1991
11. Roberts PV, Goltz MN, Mackay DM (1986) A natural gradient experiment on solute transport in a sand aquifer. 3. Retardation estimates and mass balances for organic solutes. Water Resour Res 22:2047–2058
12. Goltz MN, Roberts PV (1986) Three-dimensional solutions for solute transport in an infinite medium with mobile and immobile zones. Water Resour Res 22:1139–1148
13. Quinodoz HA, Valocchi AJ (1993) Stochastic analysis of the transport of kinetically sorbing solutes in aquifers with randomly heterogeneous hydraulic conductivity. Water Resour Res 29:3227–3240
14. Cvetkovic VD, Dagan G (1994) Transport of kinetically sorbing solute by steady random velocity in heterogeneous porous formation. J Fluid Mech 265:189–215
15. Thorbjarnarson KW, Mackay DM (1994) A forced gradient experiment on solute transport in the Borden aquifer 3. Nonequilibrium transport of the sorbing organic compounds. Water Resour Res 30:401–419
16. Fry VA, Istok JD (1994) Effects of rate-limited desorption on the feasibility of in-situ bioremediation. Water Resour Res 30:2413–2422
17. Fu MH, Mayton H, Alexander M (1994) Desorption and biodegradation of sorbed styrene in soil and aquifer solids. Environ Toxicol Chem 13:749–753
18. Aziz CE, Newell CJ, Gonzales JR, Haas PE, Clement TP, Sun Y (1999) Biochlor natural attenuation model for chlorinated solvent sites. Natural attenuation of chlorinated solvents, petroleum, hydrocarbons, and other organic compounds. Battelle Press, Columbus, OH
19. Clement TP, Johnson CD, Sun Y, Klecka GM, Bartlett C (2000) Natural attenuation of chlorinated ethene compounds: model development and field-scale application at the Dover site. J Contam Hydrol 42:113–140
20. Brusseau ML (1998) Multiprocess nonequilibrium and nonideal transport of solutes in porous media. In: Selim HM, Ma L (eds) Physical nonequilibrium in soils: modeling and application. Ann Arbor Press, Chelsea, MI

21. Fetter J (1999) Contaminant hydrogeology, 2nd edn. Prentice-Hall, Upper Saddle River, New Jersey
22. Ball WP, Roberts PV (1991) Long-term sorption of halogenated organic chemicals by aquifer material. 2. Intraparticle diffusion. Environ Sci Technol 25:1237–1249
23. Goltz, MN (1986) Three dimensional analytical modeling of diffusion limited solute transport. PhD dissertation, Stanford University, Palo Alto CA
24. Watts RJ (1997) Hazardous wastes: sources, pathways, receptors. Wiley, New York, NY
25. Wiedemeier TH, Rifai HS, Newell CJ, Wilson JT (1999) Natural attenuation of fuels and chlorinated solvents in the subsurface. Wiley, New York, NY
26. Vogel TM, Criddle CS, McCarty PL (1987) Transformations of halogenated aliphatic compounds. Environ Sci Technol 21:722–736
27. McCarty PL, Reinhard M (1993) Biological and chemical transformations of halogenated aliphatic compounds in aquatic and terrestrial environments. In: Oremland RS (ed) Biogeochemistry of global change. Chapman and Hill, New York
28. Butler EC, Hayes KF (2000) Kinetics of the transformation of halogenated aliphatic compounds by iron sulfide. Environ Sci Technol 34:422–429
29. Rittmann BE, McCarty PL (2001) Environmental biotechnology: principles and applications. McGraw-Hill, New York, NY
30. Rifai HS, Newell CJ, Gonzales JR, Wilson JT (2000) Modeling natural attenuation of fuels with BIOPLUME III. J Environ Eng 126:428–438
31. Suarez MP, Rifai, HS (1999) Biodegradation rates for fuel hydrocarbons and chlorinated solvents in groundwater. Bioremed J 3:337–362
32. Yerushalmi L, Guiot SR (1998) Kinetics of biodegradation of gasoline and its hydrocarbon constituents. Appl Microbiol Biotechnol 49:474–481
33. Bradley PM (2000) Microbial degradation of chloroethenes in groundwater systems. Hydrogeol J 8:104–111
34. Ryoo D, Shim H, Canada K, Barbieri P, Wood TK (2000) Aerobic degradation of tetrachloroethylene by toluene-*o*-xylene monooxygenase of *Pseudomonas stutzeri* OX1. Nat Biotechnol 18:775–778
35. National Research Council (NRC) (1994) Alternatives for ground water cleanup. National Academy Press, Washington, DC
36. Wilson JT, Wilson BH (1985) Biotransformation of trichloroethylene in soil. Appl Environ Microbiol 49:242–243
37. Hopkins GD, Munakata, J, Semprini L, McCarty PL (1993) Trichloroethylene concentration effects on pilot field-scale in-situ groundwater bioremediation by phenol-oxidizing microorganisms. Environ Sci Technol 27:2542–2547
38. Semprini L, McCarty PL (1992) Comparison between model simulations and field results for in-situ biorestoration of chlorinated aliphatics. Part 2. Cometabolic transformations. Ground Water 30:37–44
39. Gandhi RK, Hopkins GD, Goltz MN, Gorelick SM, McCarty PL (2002) Full-scale demonstration of in situ cometabolic biodegradation of trichloroethylene in groundwater. 2. Comprehensive analysis of field data using reactive transport modeling. Water Resour Res 38:10.1029/2001WR000380
40. Cunningham JA, Rahme H, Hopkins GD, LeBron C, Reinhard M (2001) Enhanced in situ bioremediation of BTEX-contaminated groundwater by combined injection of nitrate and sulfate. Environ Sci Technol 35:1663–1670
41. Burland SM, Edwards EA (1999) Anaerobic benzene biodegradation linked to nitrate reduction. Appl Environ Microbiol 65:529–533
42. Evans PJ, Mang DT, Young LY (1991) Degradation of toluene and *m*-xylene and transformation of *o*-xylene by denitrifying enrichment cultures. Appl Environ Microbiol 57:450–454

43. Bradley PM, Chapelle FH, Wilson JT (1998) Field and laboratory evidence for intrinsic biodegradation of vinyl chloride contamination in a Fe(III)-reducing aquifer. J Contam Hydrol 31:111–127
44. Bradley PM, Landmeyer JE, Dinicola RS (1998b) Anaerobic oxidation of [1-2-$_{14}$C] dichloroethene under Mn(IV)-reducing conditions. Appl Environ Microbiol 64:1560–1562
45. Eykhlot GR, Davenport DT (1998) Dechlorination of chloroacetanilide and herbicides alachlor and metolachlor by iron metal. Environ Sci Technol 32:1482–1487
46. Novak PJ, Christ SJ, Parkin GF (1997) Kinetics of alachlor transformation and identification of metabolites under anaerobic conditions. Water Res 31:3107–3115
47. Binks PR, Nicklin S, Bruce NC (1995) Degradation of hexahydro-1,3,5-trinitro-1,3,5-triazine (RDX) by *Stenotrophomonas maltophilia* PB1. Appl Environ Microbiol 61: 1318–1322
48. Funk SB, Roberts DJ, Crawford DL, Crawford RL (1993) Initial-phase optimization for bioremediation of munition compound contaminated soils. Appl Environ Microbiol 59:2171–2177
49. Daun G, Lenke H, Reuss M, Knackmuss HJ (1998) Biological treatment of TNT-contaminated soils. 1. Anaerobic cometabolic reduction and interaction of TNT and metabolites with soil components. Environ Sci Technol 32:1956–1963
50. Ibister JD, Anspach GL, Kitchens JF, Doyle RC (1984) Composting for decontamination of soils containing explosives. Microbiologica 7:47–73
51. McCormick NG, Cornell JH, Kaplan AM (1981) Biodegradation of hexahydro-1,3,5-trinitro-1,3,5-triazine. Appl Environ Microbiol 42:817–823
52. Hofstetter TB, Heijman CG, Haderlein SB, Holliger C, Schwarzenbach RP (1999) Complete reduction of TNT and other (poly)nitroaromatic compounds under iron-reducing subsurface conditions. Environ Sci Technol 33:1479–1487
53. Bear J (1979) Hydraulics of groundwater. McGraw-Hill, New York
54. Baetsle LH (1969) Migration of radionuclides in porous media. In: Duhamel AMF (ed) Progress in nuclear energy series XII; health physics. Pergamon Press, Elmsford, New York
55. Hunt B (1978) Dispersive sources in uniform ground-water flow. J Hydraulics Division ASCE 104(HY1):75–85
56. Domenico, PA (1987) An analytical model for multidimensional transport of a decaying contaminant species. J Hydrol 91:49–58
57. Galya DP (1987) A horizontal plane source model for ground-water transport. Ground Water 25
58. Carnahan CL, Remer JS (1984) Nonequilibrium and equilibrium sorption with a linear sorption isotherm during mass transport through an infinite porous medium: some analytical solutions. J Hydrol 73:227–258
59. Anderson MP, Woessner WW (1992) Applied ground water modeling. Academic Press, San Diego CA
60. Bedient PB, Rifai HS, Newell CJ (1999) Ground water contamination: transport and remediation, 2nd edn. Prentice-Hall, Upper Saddle River, NJ
61. Zheng C, Bennett GD (1995) Applied contaminant transport modeling: theory and practice. Van Nostrand Reinhold, New York NY
62. McDonald MG, Harbaugh AW (1988) A modular three-dimensional finite-difference ground-water flow model. Book 6. Modeling techniques, Scientific Software Group, Washington, DC
63. Zheng C (1990) MT3D, A modular three-dimensional transport model for simulation of advection, dispersion, and chemical reactions of contaminants in groundwater systems. SS Papadopulos & Associates, Rockville, MD

64. Konikow LF, Bredehoeft JD (1978) Computer model of 2-dimensional solute transport and dispersion in groundwater. In: USGS techniques of water resources investigations, book 7, chap C2. United States Geological Survey, Washington, DC
65. Konikow LF, Goode DJ, Hornberger GZ (1996) A three-dimensional method-of-characteristics solute transport model (MOC3D). Water-Resources Investigation Rep 96-4267, United States Geological Survey, Washington, DC
66. Goode DJ (1999) Age, double porosity, and simple reaction modifications for the MOC3D ground-water transport model. Water-Resources Investigation Rep 99-4041. United States Geological Survey, Washington, DC
67. Rifai HS, Newell CJ, Gonzales JR, Dendrou S, Kennedy L, Wilson J (1997) BIOPLUME III natural attenuation decision support system version 1.0 user's manual, prepared for the U.S. Air Force Center for Environmental Excellence, Brooks Air Force Base, San Antonio, TX
68. Sun Y, Peterson JN, Clement TP, Hooker BS (1996) A modular computer model for simulating natural attenuation of chlorinated organics in saturated ground-water aquifers. Proceedings, Symposium on Natural Attenuation of Chlorinated Organics in Ground Water, Dallas, TX, 11–13 Sep, EPA/540/R-96/509, US EPA, Washington, DC
69. Feng PP (2000) Modeling the effect of nonlinear and rate-limited sorption on the natural attenuation of chlorinated ethenes. MS thesis, AFIT/GEE/ENV/00M-04, Graduate School of Engineering and Management, Air Force Institute of Technology (AU), Wright-Patterson AFB OH
70. Carey GR, Van Geel PJ, Murphy JR, McBean EA, Rovers FA (1999) Modeling natural attenuation at the Plattsburgh Air Force Base. In: Alleman BC, Leeson A (eds) Natural attenuation of chlorinated solvents, petroleum hydrocarbons, and other organic compounds. Battelle Press, Columbus, OH, pp 77–82
71. Schirmer M, Molson JW, Frind EO, Barker JF (2000) Biodegradation modeling of a dissolved gasoline plume applying independent laboratory and field parameters. J Contam Hydrol 46:339–374
72. United States Environmental Protection Agency (U.S. EPA) (1999) Use of monitored natural attenuation at superfund, RCRA corrective action, and underground storage tank sites. Office of Solid Waste and Emergency Response Directive 9200.4-17P
73. Young HC, Bleckmann CA, Huang J, Reynolds DE, Goltz MN (2001) Quantitative validation of a model of chlorinated ethene natural attenuation. In: Leeson A, Kelley ME, Rafai H, Magar VS (eds) Natural attenuation of environmental contaminants. Battelle Press, Columbus, OH, pp 173–180
74. Huang J, Goltz MN (1998) A model of in situ bioremediation that includes the effect of rate-limited sorption and bioavailability, Proceedings of the 1998 Conference on Hazardous Waste Research, Snow Bird UT, 19-21 May 1998, pp 291–295
75. Kildsgaard J, Engesgaard P (2001) Numerical analysis of biological clogging in two-dimensional sand box experiments. J Contam Hydrol 50:261–285
76. Park JH, Zhao X, Voice TC (2001) Biodegradation of non-desorbable naphthalene in soils. Environ Sci Technol 35:2734–2740
77. Voice TC, Zhao X, Maraqa MA (1997) Desorption/biodegradation interactions in laboratory soil columns. American Chemical Society Division of Environmental Chemistry, Preprints of Extended Abstracts 37:245–246

Handb Environ Chem Vol. 5, Part F, Vol. 3 (2005): 65–96
DOI 10.1007/b11442

Mathematical Methods for Hydrologic Inversion: The Case of Pollution Source Identification

Amvrossios C. Bagtzoglou[1] (✉) · Juliana Atmadja[2]

[1] Department of Civil and Environmental Engineering, 261 Glenbrook Road, Unit 2037, University of Connecticut, Storrs, Connecticut 06269-2037, USA
acb@engr.uconn.edu

[2] TAMS, an Earth Tech Company, 300 Broadacres Drive, Bloomfield, NJ 07003, USA
Juliana.Atmadja@earthtech.com

1	**Introduction**	66
2	**Groundwater Contaminant Transport**	69
2.1	Advection-Dispersion Equation	70
2.2	Pollution Source Identification	70
3	**Methods for Source Identification**	73
3.1	Heat Transport Inversion Methods	73
3.2	Contaminant Transport Inversion Methods	75
3.2.1	Optimization Approaches	75
3.2.2	Probabilistic and Geostatistical Simulation Approaches	78
3.2.3	Analytical Solution and Regression Approaches	83
3.2.4	Direct Approaches	85
4	**Conclusions**	93
References		93

Abstract The reliable assessment of hazards or risks arising from groundwater contamination problems and the design of efficient and effective techniques to mitigate these problems require the capability to predict the behavior of chemical contaminants in flowing water. To choose an appropriate remediation strategy, knowledge of the contaminant release source and time release history becomes pertinent. With more and more contamination sites being detected nowadays, it is almost impossible to perform exhaustive drilling, testing, and chemical fingerprint analysis every time, especially in the case of pollution being generated by highway construction and repair materials. Moreover, most of the time, chemical finger printing, state and federal agency records, and private parties' history records of handling hazardous substances are not sufficient to allow a unique solution for the timing of source releases. The purpose of this chapter is to present and review mathematical methods that have been developed during the past 15 years to perform hydrologic inversion and specifically to identify the contaminant source location and time-release history.

Keywords Inverse problem · Hydrologic modeling · Contaminant transport · Heat conduction · Groundwater pollution

List of Abbreviations

ADE	Advection-Dispersion Equation
AFB	Air Force Base
BBE	Backward beam equation
CERCLA	Comprehensive Environmental Response, Compensation and Liability Act
C&R	Construction and repair
DNAPL	Dense non-aqueous phase liquids
EPA	Environmental Protection Agency
GW	Groundwater
LUST	Leaking underground storage tanks
MJBBE	Marching jury backward beam equation
MRE	Minimum relative entropy
NLS	Non-linear squares
PCE	Tetrachloroethylene
PDF	Probability density function
PGA	Progressive genetic algorithm
QR	Quasi-reversibility
QBVP	Quasi-boundary-value-point
TCE	Trichloroethylene
TR	Tikhonov regularization
SW	Solid waste
1D	One-dimensional
2D	Two-dimensional
3D	Three-dimensional

1 Introduction

As the world's population continues to grow, the demand for fresh water will certainly increase. Seventy five percent of the Earth's surface is covered with water. Out of this 75%, almost 97% of the world's water is salty and not readily drinkable. The other 2% is locked as solids, in the form of ice caps and glaciers, leaving us with about 1% of freshwater that is available for all humanity's need. This small amount of freshwater can be found in the form of surface water and groundwater. Of the 1% of freshwater available, 96% is in the form of groundwater. From 1940 to 1990, withdrawals of fresh water from rivers, lakes, reservoirs, and other sources have been augmented fourfold [28].

Among the freshwater available for use, groundwater amounts to about half of the U.S. population's source of drinking water. However, quantity is not the only issue of concern; the quality of the drinking water is also a problem since this vital resource is vulnerable to contamination. With the advances of technology, more and more human activities can easily pollute the water system. Water pollutants may originate from a point source or a non-point source. A point source is characterized by the presence of identifiable, small-scale sources while the non-point problem refers to larger-scale, more widespread contamination originating from many smaller sources. The United States Environmental Protection Agency (EPA) reported that as of 1996, about 40% of the nation's surveyed rivers, lakes, and estuaries are too polluted for basic uses, such as fishing and swimming [64]. In most cases, pollution in the surface water is caused by runoff and groundwater seepage. In the U.S. EPA 305(b) Report [64] 37 states have reported that they found potential sources of groundwater contamination.

Major sources of point-source pollution are Leaking Underground Storage Tanks (LUSTs), landfills, septic systems, and hazardous waste sites. Among the major non-point sources, pesticides, leaks or spills of industrial chemicals at manufacturing facilities, runoff of salt and other chemicals from roads and highways, and fertilizers on agricultural land are high on the list. As of March 1996, more than 300,000 releases from LUSTs had been confirmed [65]. Fertilizers, pesticides, and agricultural chemical facilities also contribute to the contamination in the water as non-point sources. Most of water contamination cases occur in highly developed areas, agricultural areas, and industrial zones. In addition to affecting human health, pollution is also detrimental to natural resources and ecosystems. Once a contaminant is detected, an effort should be, and is being made, to clean up the groundwater system. In 1980, the U.S. Congress passed the Comprehensive Environmental Response, Compensation and Liability Act (CERCLA) along with an accompanying legislation called the Superfund Tax Act. The Superfund acts as a "bank" to create a fund for cleaning up contaminated sites. Besides financing the clean up process, the Superfund is also responsible to locate and investigate, and clean up the worst sites nationwide. The Superfund cleanup process begins with the site discovery and/or notification to EPA. Once the site is identified, it is entered into the Comprehensive Environment Response, Compensation, and Liability Information System. Due to limited resources available for the cost recovery program, the EPA must set priorities and select plan of actions in a manner and at a time that will provide for the maximum return to the Fund [66]. After the priority selection, the clean up process begins according to the list. This procedure is followed by recovering the cost from the identified responsible private parties. Alternatively, the EPA can order a private party to perform the clean up or obtain a voluntary agreement from such private party to perform clean up [50]. In case a private party is not identified, the money comes from the Superfund "bank", which essentially is the tax-payer's money. The legislation has not been an unqualified success in assigning the liability.

The cost of cleanup is staggering. This can be seen by the fact that in 1994, the National Academy of Sciences estimated that over a trillion dollars, or approximately $4000 per person in the U.S., would be spent in the next 30 years on clean up of contaminated soil and groundwater [67]. In many cases, it is hard to find out which companies or parties are responsible for the contamination due to lack of tools to identify the pollution source. Chemical finger printing, state and federal agency records, and private parties' history records of handling hazardous substances are seldom sufficient to allow a unique solution for the location of sources and the timing of source releases. Therefore, there exists a need for tools that can reconstruct the plume spatial and temporal history [50].

Recycled hazardous solid wastes are nowadays used as highway construction and repair (C&R) materials. Such practice has been identified as a non-point source of environmental pollution since constituents of highway materials can migrate from roadways to the surrounding environment, presenting a potential pollution source. Parallel to the increased utilization of C&R materials, solid waste (SW) materials have also started being used, thus raising additional concerns and providing the impetus for evaluating the potential for environmental contamination due to leachates from highway C&R and SW materials. There is a clear need to integrate and unify testing and evaluation approaches that will: (i) allow greater understanding of the fundamental leaching behavior of such SW and C&R materials, (ii) help model the fate, transport and toxicity of such SW and C&R materials, and (iii) identify where the major source, or series of major sources, of contamination happen to be located or estimate the contaminant time-release history so that removal or remediation can be initiated.

Based on the needs discussed above, it is clear that the reliable assessment of hazards or risks arising from groundwater contamination problems and the design of efficient and effective techniques to mitigate these problems require the capability to predict the behavior of chemical contaminants in flowing water. To choose an appropriate remediation strategy, knowledge of the contaminant release source and time release history becomes pertinent. With more and more contamination sites being detected nowadays, it is almost impossible to perform exhaustive drilling, testing, and chemical fingerprint analysis every time, especially in the case of highway C&R materials. Moreover, most of the time, chemical finger printing, state and federal agency records, and private parties' history records of handling hazardous substances are not sufficient to allow a unique solution for the timing of source releases.

Quantitative predictions of contaminant movement can be made only if we understand the processes controlling transport, advection, hydrodynamic dispersion, chemical, physical, and biological reactions that affect solute concentrations in the water. To predict the behavior of contaminants, the effects of each of these influences must be adequately represented in a model or group of models. One area of contaminant behavior that has been researched for quite some time is the chemical behavior aspect. Chemical finger printing has been a major element in environmental investigation to identify the contaminants in groundwater [26, 45, 54]. A recent paper by Morrison [47] presented an extensive literature review of commonly used environmental forensic methods for age dating and source identification. In this paper, Morrison presented an overview of commonly used environmental forensic techniques and their possible applications so the readers can decide which technique is most appropriate for their cases. A more in-depth review of these techniques can be found in Morrison [48, 49]. Finger printing alone is not always sufficient to provide answers to questions of source and responsibility [62]. In most cases, additional information such as the release and spill history of the contaminant, chemical parent product types from which contaminants have originated, and site historical records are needed. Furthermore, by accurately identifying pollution sources, the complex and lengthy process of remediation that is often initiated by finding the contamination source can be shortened.

2 Groundwater Contaminant Transport

Contaminants dissolved in the subsurface environment are transported by three processes: advection, mechanical dispersion, and molecular diffusion. Advection resorts to the contaminant being carried along with the flow of subsurface water. This process is a result of the large-scale gradients in fluid energy (head) and it is the most significant mass transport process. The velocity of the water is described by the average speed of the water movement through the pores of the soil. Mechanical dispersion and molecular diffusion collectively are referred to as hydrodynamic dispersion. Mechanical dispersion is the main process that causes the contaminant to spread out and be diluted. The contaminant spreads due to the spatial variation of flow paths and the variation of velocity in the groundwater movement. The amount of mechanical dispersion increases with the heterogeneity of the aquifer and the scales involved. Molecular diffusion is the process in which the contaminants move from high concentration areas to low concentration areas due to concentration gradients according to Fick's Law. This process can take place in the absence of groundwater flow. All three mechanisms mentioned above are causing the contaminant to spread in the direction of flow both longitudinally and transversally (Fig. 1).

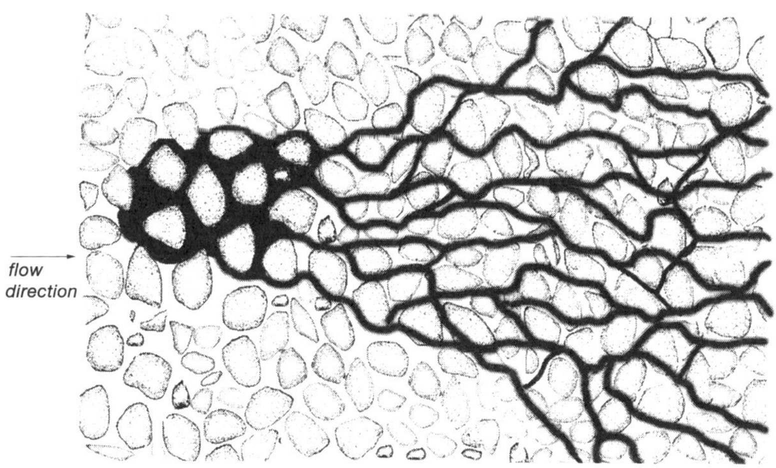

Fig. 1 Contaminant transported in porous media

In the process of being transported, many contaminants react with other compounds in the solution or the soil grains. The contaminant can be adsorbed onto or desorbed off the solid matrix. These mechanisms are known as reactions. Various types of reaction, ranging from simple sorption/desorption to radioactive decay, exist in contaminant transport. Reactive contaminant transport is an area of active and intense research. In this chapter, only non-reactive contaminant transport will be considered.

2.1 Advection-Dispersion Equation

Non-reactive contaminant transport may be represented by the Advection-Dispersion Equation (ADE) in three-dimensional (3D) form:

$$\frac{\partial C}{\partial t} = \nabla [D \cdot \nabla C] - \nabla [v\, C] \tag{1}$$

where ∇ is the gradient operator, C is the solute concentration, D is the dispersion coefficient tensor, v is the transport velocity vector, and t is time.

To understand the behavior of the movement of the contaminant in groundwater, people solve Eq. (1) forward in time. In solving this equation forward in time, one assumes that the plume is originated from somewhere and will travel through the porous media due to advection and dispersion. The conventional procedure to solve Eq. (1) is to use finite difference or finite element methods. For simple cases, closed-form solutions exist. Quantitative descriptions of the processes forward in time are well understood. Multidimensional models of these processes have been used successfully in practice [50]. Numerical solute transport models were first developed about 25 years ago. When properly applied, these models can provide useful information about transport processes and can assist in the design of remedial programs.

2.2 Pollution Source Identification

One-dimensional (1D) heterogeneous non-reactive contaminant transport in a semi-infinite domain, described by the ADE is

$$\frac{\partial C}{\partial t} = \frac{\partial}{\partial x}\left[D(x)\frac{\partial C}{\partial x}\right] - \frac{\partial}{\partial x}[u(x)\, C] \tag{2}$$

$$C(x_1, t) = C_{in}(t) \qquad 0 \le t \le T_{obs} \tag{3}$$

$$C(\infty, t) = 0 \qquad 0 \le t \le T_{obs} \tag{4}$$

$$C(x, T_{obs}) = C_T(x) \qquad 0 \le x \le \infty \tag{5}$$

where C is the solute concentration, $D(x)$ is the dispersion coefficient, $u(x)$ is the transport velocity in the x-direction, x is distance, and t is time. In the pollution source location identification, the source (x_1) information is not known, but measurements of the spatial distribution of the plume are given at time T_{obs}. For the release history reconstruction, usually the source location is assumed to be known, but the contaminant source function, $C_{in}(t)$, is unknown.

Finding the source location and the time history of the solute in groundwater can be categorized as a problem of time inversion. This means that we have to solve the governing equations backward in time. Modeling contaminant transport using reverse time is an ill-posed problem since the process, being dispersive is irreversible. Because of this ill-posedness, the problems have discontinuous dependence on data and are sensitive to the errors in data.

A problem is categorized as a well-posed problem if (a) the solution exists, (b) the solution is unique, and (c) the solution is stable [63]. Problems that do not satisfy these criteria are called ill-posed. For the groundwater contamination problem, the plume has to have originated from someplace; therefore, physically, the plume exists. However, in mathematically rigorous terms, the fact that we have a present day plume concentration, does not necessarily mean that we satisfy the existence criterion. The solution exists only when we have perfect and consistent model and data that satisfy extremely restrictive conditions. Satisfying the stability criterion is a difficult task to accomplish since numerical schemes which are usually implemented as a marching procedure are unstable for negative time steps, and make it impossible to solve contaminant

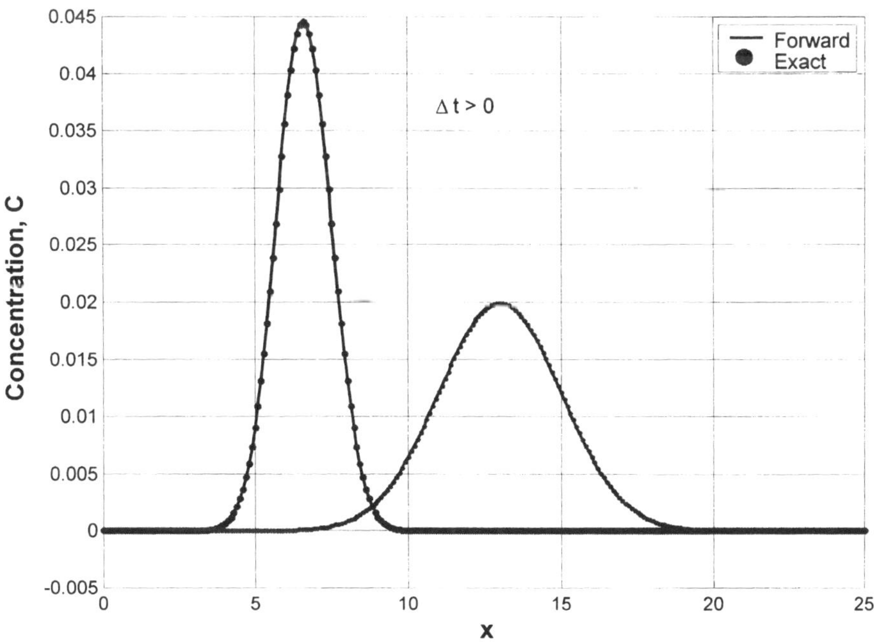

Fig. 2 Finite difference solution of the ADE for forward simulation with $\Delta t>0$

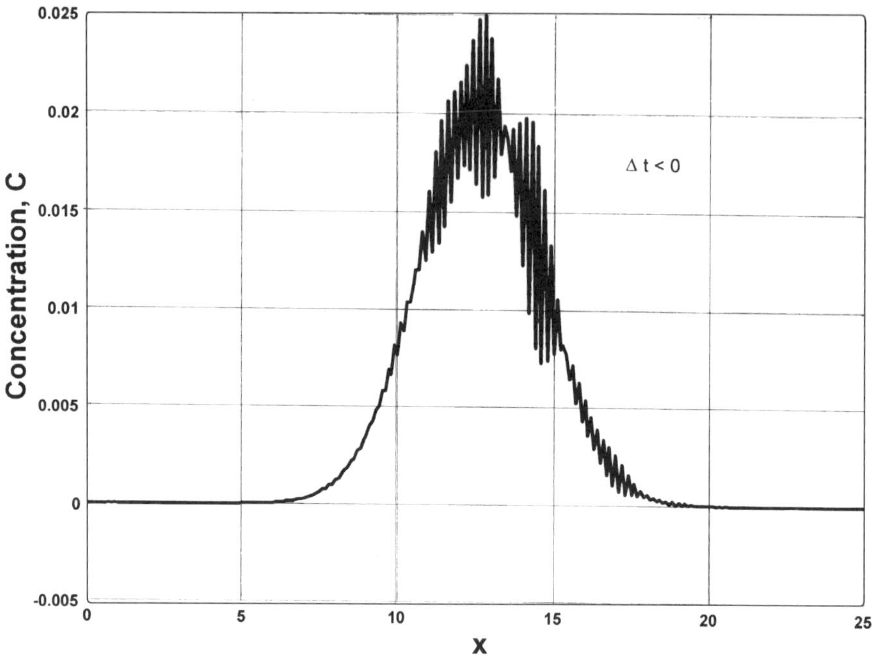

Fig. 3 Finite difference solution of the ADE for backward simulation with $\Delta t<0$

transport problems backward in time. This can be seen from a numerical example of the propagation of a rectangular pulse of solute. Using a Crank-Nicolson finite difference scheme, the plume is propagated forward in time for 1600 time steps, from time t=0.4 to t=2 with Δx=0.1 and Δt=0.001. The numerical results are in good agreement with the exact solutions (Fig. 2). On the other hand, starting at t=2 and using the same scheme with Δt=−0.001 leads to severe instability after just 89 time steps (Fig. 3).

As for the non-uniqueness of the solution, there is no method that can bypass this inherent problem. In inverse problems, one of the common practices to overcome the stability and non-uniqueness criteria is to make assumptions about the nature of the unknown function so that the finite amount of data in observations is sufficient to determine that function. This can be achieved by converting the ill-posed problem to a properly posed one by stabilization or regularization methods. In the case of groundwater pollution source identification, most of the time we have additional information such as potential release sites and chemical fingerprints of the plume that can help us in the task at hand.

3 Methods for Source Identification

3.1 Heat Transport Inversion Methods

The physical and mathematical models of heat and mass transfer are similar. The governing equation for the heat conduction problem is similar to Eq. (1), without the advective or convective term on the right hand side of the equation. Research in the inverse heat transfer area has been going on for about three decades and numerous methods have been proposed. Therefore, it would be impossible to mention all of them in this chapter. An extensive review has been presented by Beck et al. [15] and Hensel [32]. Early methods included approaches using linear programming by Cannon [20], who presented numerical inversion of the heat equation in his paper, Lattes and Lions [37], who proposed the Quasi-Reversibility (QR) method to solve heat conduction problems, and Ewing [29], who used the Sobolev equations to approximate the backward solution of parabolic equations. The QR method has also been used for the solution of ill-posed heat conduction problems in a Quasi-Boundary-Value-Problem (QBVP) variant of the original formulation [25]. The QR method has also been applied to contaminant transport problems. Therefore, this method will be described in more details later. The conjugate gradient method with minimization procedures was also used for the linear and non-linear inverse heat transfer problem [16, 30, 35, 36]. The conjugate gradient method, in general, is used to solve linear algebraic equations simultaneously:

$$Ay = g \tag{6}$$

where A is an $N \times N$ matrix, y is an $N \times 1$ vector of unknowns, and g is an $N \times 1$ vector of constants. The solution of Eq. (6) may or may not exist depending whether or not g is in the range of A, $\mathfrak{R}(A)$. One can obtain a best approximate solution by minimizing the quadratic functional

$$\|Ay - g\|^2 \tag{7}$$

Solution for Eq. (7) always exists for all $g \in G$, where G is a Hilbert space, if $\mathfrak{R}(A)$ is closed.

Huang and Ozisik [33] solved the inverse forced convection problem that is formulated as follows:

$$C_k \frac{\partial T(x,y)}{\partial y^2} = u(y)\rho C_p \frac{\partial T(x,y)}{\partial x} \qquad \text{in } 0 < y < L \quad 0 < x < b \tag{8}$$

$$\frac{\partial T(x,0)}{\partial y} = 0 \qquad \text{at } y = 0 \tag{9}$$

$$C_k \frac{\partial T(x,L)}{\partial y} = q(x) \qquad \text{at } y = L \tag{10}$$

$$T(0,y) = T_0 \qquad \text{at } x = 0 \tag{11}$$

where C_k is thermal conductivity, $T(x,y)$ is the temperature distribution, $u(y)$ is velocity, ρ is density, C_p is heat capacity and $q(x)$ is the wall heat flux. For the inverse problem, $q(x)$ is unknown and the temperature measurements were taken at S different positions, x_i, with $i=1,\ldots,S$, at a distance $y=y_1$ from the lower wall. Huang and Ozisik employed the combination of regular and modified conjugate gradient methods to determine the unknown wall heat flux for laminar flow inside a parallel plate duct. The adjoint expression for Eq. (8) is

$$J(q) = \int_0^b [T(x,y_1) - Y(x,y_1)]^2\,dx + \int_0^b\int_0^L \lambda(x,y)\left[C_k\frac{\partial T}{\partial y^2} - u(y)\,\rho C_p\frac{\partial T}{\partial x}\right]dy\,dx \tag{12}$$

Perturbing $T(x,y)$ by $\Delta T(x,y)$ and $q(x)$ by $\Delta q(x)$ in Eq. (12), the variation of the functional ΔJ is determined as

$$\Delta J = \int_0^b J'(q)\Delta q(x)\,dx \tag{13}$$

where $J'(q)=\lambda(x,L)$ is the gradient equation that relates the gradient of the functional $J(x)$ to the adjoint function $\lambda(x,L)$.

The iterative procedure for the solution of the inverse problem by the regular conjugate gradient method was performed assuming the functions $T(x,y)$, $\Delta T(x,y)$, $\lambda(x,y)$, and $J'(x)$ are available at the n-th iteration. The wall heat flux, $q(x)$, at step $n+1$ is obtained from

$$q^{n+1} = q^n - \beta^n P^n \qquad n = 0,1,2,\ldots \tag{14}$$

where P^n is the direction of descent, determined from

$$P^n(q) = J'^n(q) + \gamma^n P^{n-1}(q) \tag{15}$$

and the conjugate coefficient γ^n is computed as follows:

$$\gamma^n = \frac{\int_{x=0}^{b} [J'^n(x)]^2\,dx}{\int_{x=0}^{b} [J'^{n-1}(x)]^2\,dx} \qquad \text{with } \gamma^0 = 0 \tag{16}$$

The modified conjugate gradient method was developed to alleviate the effect of the final position on convergence. In the modified method, the direction of the descent $P^n \equiv P^n(q)$ is related to the derivative of the direction of descent $R^n \equiv R^n\,(dq/dx)$ by the following relation:

$$P^n = \int R^n(\xi)\,d\xi \tag{17}$$

and R^n is determined from

$$R^n\left(\frac{dq}{dx}\right) = J'^n\left(\frac{dq}{dx}\right) + \gamma^n R^{n-1}\left(\frac{dq}{dx}\right) \tag{18}$$

While Huang and Ozisik solved the spacewise variation of wall heat flux for laminar forced convection problem, Silva Neto and Ozisik [57] used the conjugate gradient method and the adjoint equation simultaneously to solve for the timewise-varying strength of a two plane heat source.

Different methods to eliminate the ill-posedness of the problem were developed, such as Tikhonov Regularization (TR) and stabilization [3, 4, 38, 63]. Ames and coworkers [5] presented a comparison of regularization approaches for the backward heat equation problem. The TR approach has been used extensively in various fields of applications, one of which is the contaminant source identification. Therefore, this method will be discussed in details later. A finite element approach without stabilization procedures was introduced by Bayo et al. [14]. The computational procedure was made efficient by using the fast Fourier transform. Although Bayo and co-workers presented the application of the inverse heat conduction problem for the finite element method in their paper, the method can be used for finite differences or any other technique for spatial discretization. Marquardt and Auracher [46] introduced the observer-based technique for solving the inverse heat conduction problem. Carasso [23] constructed consistent space marching difference schemes to solve the nonlinear inverse heat conduction problem that blow-up more slowly than the counterpart analytical problem. Carasso examined various combinations of space and time differencing that lead to a total of 18 different algorithms.

3.2 Contaminant Transport Inversion Methods

3.2.1 Optimization Approaches

For the past 20 years, several attempts have been made to solve the ADE to identify the pollution source. One of the early methods to backtrack the pollution source location was to run forward simulations and check the solutions with the measured/current spatial data observed. Due to the non-uniqueness of the solution and the infinite number of plausible combinations, one needs to follow an optimization method to obtain the best fitted solution. Among the first people to tackle pollution source identification problems using optimization approaches are Gorelick et al. [31]. They formulated the groundwater pollution source identification problem as forwardtime simulations in conjunction with an optimization model using linear programming and multiple regressions. In their approach, they incorporated the solute transport model as a series of constraints in the form of a concentration response matrix. Two examples were used: (a) a steady state case, where the method is used to locate the unknown pollutant sources constructed from concentration data collected at few locations by means of a subsurface pipe that lies in the unsaturated zone with a selected leak location, and (b) a transient case, where the concentration history was used to identify the source schedule for a complex two-dimensional (2D) system in which several wells are responsible for the observed pollution. The ADE for steady state case used was a 2D transport with linear decay of a single dissolved chemical constituent in saturated porous media, described as follows:

$$\frac{\partial}{\partial x_i}\left(D\frac{\partial C}{\partial x_j}\right)-\frac{\partial}{\partial x_i}(Cv_i)-\lambda C=\frac{C'W}{\phi b} \qquad i,j=1,2 \tag{19}$$

where b is the saturated aquifer thickness, ϕ is the effective aquifer porosity, W is the volume flux (source) per unit area, and λ is the first order kinetic decay rate. To determine the pipe leak locations and leak magnitudes, one needs to minimize the sum of the absolute normalized residuals. The normalized residual r_n is described as

$$r_n=\frac{C_s-C_{obs}}{C^*} \tag{20}$$

where C_s is the simulated concentration, C_{obs} is the observed concentration and C^* is the estimated true concentration. The linear programming objective is

$$\text{Min } z=\{e^t\}\{|r_n|\} \tag{21}$$

where $\{|r_n|\}$ is a vector that consists of the absolute values of the normalized residuals r_{ni} corresponding to the i-th simulated-observed concentration pair and $\{e^t\}$ is the identity vector.

In the steady state case example, because the data were sparse and there were more unknowns than constraining equations, an additional restriction to the linear programming method was needed. This was done by utilizing a mixed integer-programming model such as in a stepwise multiple regression solution. In the multiple regression method, the identification problem was formulated as follows:

$$\text{Min } Z'=\{e^t\}\{r_n^2\} \tag{22}$$

subject to:

$$[R]\{q\}-[I]\{r_n\}=\{e\} \tag{23}$$

where $[R]$ is the rectangular matrix of m constraints, n unknowns, and containing the concentration response of information for both chloride and tritium and $\{q\}$ is the vector of unknown source flux magnitude. While using the multiple regressions, the results displayed spurious negative values that detracted from the true solution. For the transient case, both methods identified the pollution source and the disposal episodes, but contained some errors in determining the disposal flux magnitude. In both models, Gorelick and coworkers assumed no uncertainty in the physical parameters of the aquifer. The method is restricted to cases where data are available in the form of breakthrough curves.

Another optimization approach was followed by Wagner [68]. Wagner developed a methodology for performing simultaneous model parameter estimation and source characterization, in which he used an inverse model as a non-linear maximum likelihood estimation problem. The hydrogeologic and source parameters were estimated based on hydraulic head and contaminant concentration measurements. In essence, this method is minimizing the following:

$$G = \frac{N}{2}\ln(2\pi) + \frac{1}{2}\ln|C| + \frac{1}{2\sigma_h^2}G_h + \frac{1}{2\sigma_c^2}G_c \tag{24}$$

where

$$G_h = [h^* - h(p)]V_h^{-1}[h^* - h(p)] \tag{25}$$

$$G_c = [c^* - c(p)]V_c^{-1}[c^* - c(p)] \tag{26}$$

where h^* is the hydraulic head measurements, c^* is the contaminant concentration measurements, $h(p)$ and $c(p)$ are the modeled heads and concentrations, respectively. C is the covariance, represented by

$$C = \begin{bmatrix} C_h & 0 \\ 0 & C_c \end{bmatrix} \tag{27}$$

where

$$C_h = \sigma_h^2 V_h \tag{28}$$

$$C_c = \sigma_c^2 V_c \tag{29}$$

and where V_h and V_c are symmetric positive definite matrices and σ_h^2 and σ_c^2 are positive scalars.

Wagner assumed a steady confined groundwater flow and transient non-reactive single-species solute transport for the example problem. His method is capable of providing accurate and reliable estimates of the unknown model parameters, which are within 10% of the true parameter values. While considering the magnitude of disposal flux as unknown, the method became less accurate. In this case, the estimated parameters are within 30% of the true values and the estimated contamination flux is 20% greater than the true value. Wagner did go a step further, that is assuming an unknown contaminant release history. Assuming that the aquifer parameters, exact time span, and magnitude of disposal flux are unknown, the method gives fairly accurate results, however, the standard deviations for the estimated aquifer parameters increased by as much as 37%. The contaminant flux estimate remains the same as before. The method is also capable of quantifying the history of the disposal accurately. The last model used by Wagner was identifying the contaminant disposal source together with the parameters being estimated. For this example, it was assumed that there existed two potential disposal areas. The method estimated the parameters fairly accurately, but the standard deviations increased up to 55% compared to the case of unknown disposal history. As for the source characterization, the maximum likelihood estimated the true source location and the true disposal rate fairly accurately. The contaminant flux estimate at the true location was approximately 6% greater than the true disposal flux.

Along the same line of work, Mahar and Datta [42, 44] developed a methodology that combines the concepts of optimal identification of pollutant source with the optimal design of groundwater quality monitoring network for an efficient identification process. They applied their method to a hypothetical 2D homogeneous, isotropic, and saturated aquifer with a conservative pollutant plume. The first step in their methodology is to utilize an optimization model for preliminary identification of an unknown pollution source based on observation data. The next step is to simulate different realizations of pollution plumes using perturbed sources. The solution of the simulation model for a given set of source fluxes represents a probable realization of the pollutant plume. Once the realizations are obtained, integer-programming is used to determine the optimal locations of the quality monitoring wells. Finally, the concentration measurements made at the designed monitoring well locations are used in the non-linear optimization model to obtain a more accurate estimation of the sources. The objective function of the nonlinear source identification model proposed by Mahar and Datta is

$$\text{Min } F = \sum_{n=1}^{nk} \sum_{iob=1}^{nob} (cest_{iob}^{n} - cobs_{iob}^{n})^2 \, w_{iob}^{n} \tag{30}$$

where $cest_{iob}^{n}$ is the concentration estimated by the identification model at observation well location iob and at the end of time period n, nk is the total number of concentration observation time periods, nob is the total number of observation wells, $cobs_{iob}^{n}$ is the observed concentration at well iob at the end of time period n, and w_{iob}^{n} is the weight corresponding to observation location iob at the end of time period n, and is specified as

$$w_{iob}^{n} = \frac{1}{(cobs_{iob}^{n} + \eta^2)} \tag{31}$$

where η is a constant. Mahar and Datta [43] also used nonlinear optimization to estimate the magnitude, location and duration of groundwater pollution sources under transient conditions. Recently, Aral et al. [6] applied the Progressive Genetic Algorithm (PGA) to the problem of source location and release history identification. The PGA is a combinatorial optimization approach that is coupled easily with the nonlinear optimization model formulation used for source identification.

3.2.2
Probabilistic and Geostatistical Simulation Approaches

Bagtzoglou [10] and Bagtzoglou et al. [11, 12] are among the first to attempt solving the ADE backwards in time without relying on optimization approaches. In their work, they modeled the reversed time transport equation using the

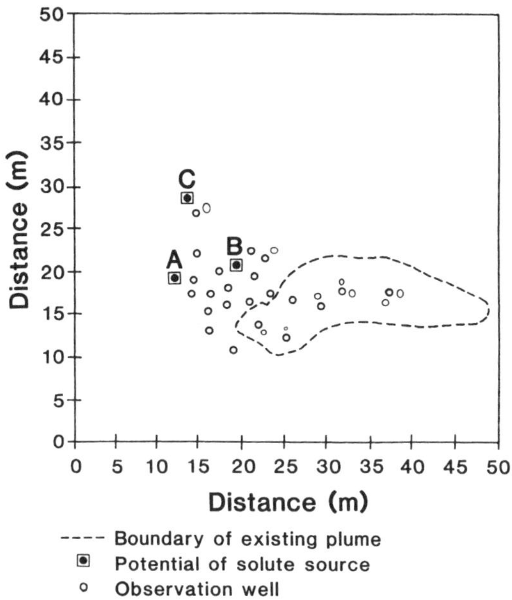

Fig. 4 Hypothetical aquifer system for source identification with conditional hydraulic conductivity fields

random walk particle method. In this approach, the advective part of the transport model is reversed while the dispersive part is left unchanged. They presented a probabilistic framework to identify solute sources in heterogeneous media. The example problem used for this study was a heterogeneous aquifer system with three potential sources of contamination (Fig. 4).

In all simulations the spill incidents were assumed to be instantaneous and occurring simultaneously. A particle representation of the concentration field (at the present time) was evaluated. Assuming that the hydraulic conductivity field is known with certainty, the velocity field was reversed. Repeated reversed time solute transport realizations were conducted and the first two moments of the concentration probability density function (*pdf*) were obtained. Using geostatistical techniques, Bagtzoglou and co-workers successfully assessed the relative importance of each potential source using spatial (Fig. 5a) and temporal (Fig. 5b) variations for the identification *pdf*. Their approach was also extended to cases where the hydraulic conductivity field is not known with certainty with the help of conditional geostatistical simulations. Figure 6 depicts the temporal variation of the identification *pdf* values for a source identification scenario involving only one source at point A of Fig. 4. The aquifer heterogeneity is σ_Y=1.0, where Y=ln(K) and K is the hydraulic conductivity of the aquifer material. The effect the number of conditioning points has on the reliability of the identification is evident when comparing Fig. 6a (20 points) with 6b (60 points). It is worth mentioning that for the scenario discussed here

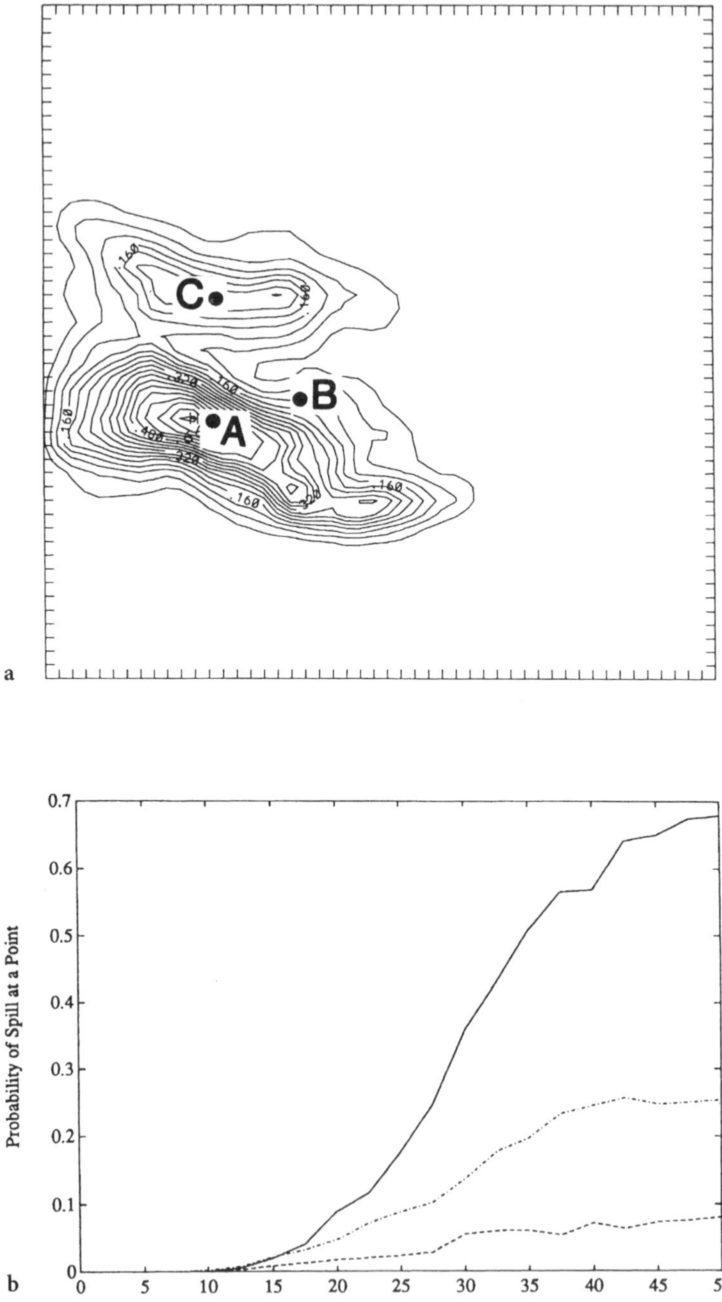

Fig. 5a, b Identification *pdf* values for a scenario with only two sources (A and C) and a σ_Y=1.0 for perfectly known hydraulic conductivity field: **a** spatial variation at t=50; **b** temporal variation for all possible sources

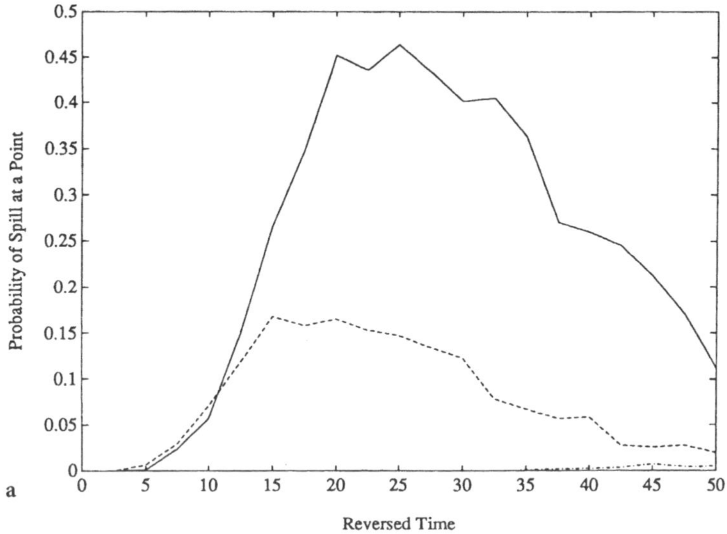

a

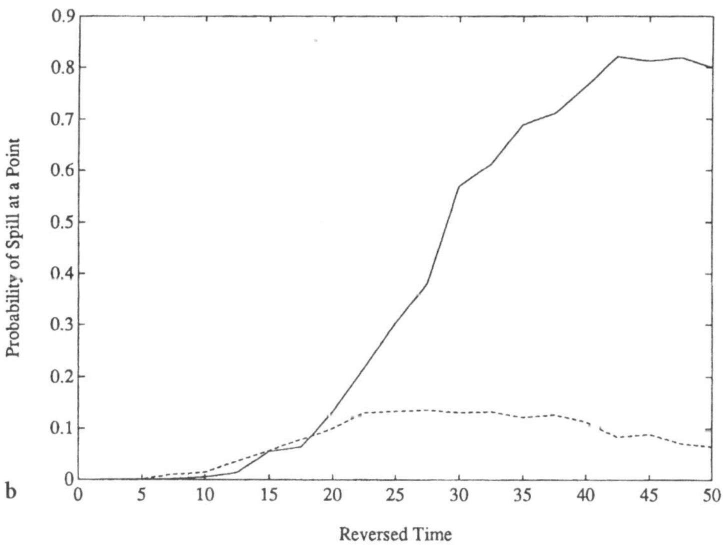

b

Fig. 6a, b Temporal variation of identification *pdf* values for a scenario with only one source (A) and a σ_Y=1.0 for: **a** 20 conditioning points; **b** 60 conditioning points for the hydraulic conductivity field

the *pdf* plot for source A (solid line) should reach 1.0 at 50 units of reversed time and there should exist no contribution from source B (dashed line).

In the line of probabilistic approaches, Wilson and Liu [69] solved the transport equation using stochastic differential equations backwards in time. Like in the method developed by Bagtzoglou and co-workers, Wilson and Liu kept the dispersion part positive and reversed the advection part. They provided two types of maps, namely travel time probability and location probability. The travel time probability map was obtained from the normalized flux concentration, given by

$$f(t|x_o=0) = \frac{QC^f(x=0,t)}{\int_0^\infty QC^f(x=0,t')\,dt'} = \frac{v}{\sqrt{\pi Dt}}\exp\left[\frac{-(x_o-vt)^2}{4Dt}\right] - \frac{v^2}{2D}\exp\left[\frac{vx_o}{D}\right] erfc\left[\frac{x_o+vt}{\sqrt{4Dt}}\right] \tag{32}$$

where $f(t|x_o)$ is the travel time *pdf* for a given source location (x_o), and Q is the pumping rate. The source location *pdf* describes the probability of the arrival of the non-reacting contaminant particle at the pumping well and is given by

$$f(x'|t,x=0) = \frac{f(t|x',x=0)\,f(x')}{\int_0^\infty f(t|x',x=0)\,f(x')\,dx'} \tag{33}$$

Wilson and Liu showed that both location and travel time probabilities can be calculated directly, using a backward-in-time version of traditional continuum advection-dispersion modeling. In addition, they claimed that by choosing the boundary conditions properly, the method can be readily generalized to include linear adsorption with kinetic effects and 1st order decay. An extension of their study for a 2D heterogeneous aquifer was reported in Liu and Wilson [39]. The results for travel time probability are in very close agreement with the simulation results from traditional forward-in-time methods.

Results from this model were verified by Neupauer and Wilson [51] using the adjoint method. In this method, the forward governing equation, with concentration as the dependent variable, is replaced by the adjoint equation, with the adjoint state as the dependent variable. They showed that backward-in-time location and travel time probabilities are adjoint states of the forward-in-time resident concentration. In this and the follow-up paper, Neupauer and Wilson [51, 52] presented the adjoint method as a formal framework for obtaining the backward-in-time probabilities for multidimensional problems and more complex domain geometries.

Snodgrass and Kitanidis [61] also used a probabilistic approach combining Bayesian theory and geostatistical techniques. In their method, the source function to be estimated is discretized into components that are assigned a known stochastic structure with unknown stochastic parameters. The method incorporates uncertainty in contaminant concentration. The unknown function, for example the release history, is represented by a random process. The estimation problem solved by Snodgrass and Kitanidis is described as

$$z = Hs + \varepsilon \tag{34}$$

where z is a vector of observations, H is a known matrix, s is the 'state vector' obtained from the discretized of the unknown function to be estimated, and ε is the measurement error vector. The vector s was then estimated by maximizing its posterior probability. Snodgrass and Kitanidis' method is an improvement from some other methods in that the solutions are more general and make no blind assumptions about the nature and structure of the unknown source function. Limitation to this approach is that the location of the potential source must be known a priori.

3.2.3 Analytical Solution and Regression Approaches

Sidauruk et al. [56] presented an inverse method based on analytical solutions of contaminant transport problems. Using their method, one only needs a well planned set of concentrations, and no other prior information. The method provides a complete estimate of the dispersion coefficient, flow velocity, amount of pollutant, its initial location, and time origin. However, since this method is based on an analytical solution, it works only for very limited cases. The method is applicable only to homogeneous aquifers with simple geometries and flow conditions. In their paper, Sidauruk and co-workers looked at two simple cases of instantaneous and continuous point releases in a 2D uniform groundwater flow field. They took the exact solutions and obtained the linear relations between the logarithm of concentration and certain combination of parameters. For the wells along the x-axis (y=0), the logarithm of the concentration is described as

$$\ln c = m_x \mathrm{X} + b_x \tag{35}$$

and for the wells along the y-axis (x=0)

$$\ln c = m_y \mathrm{Y} + b_y \tag{36}$$

where

$$\mathrm{X} = (x - x_m)^2 \tag{37}$$

$$x_m = x_o - vt_o \tag{38}$$

$$\mathrm{Y} = (y - y_o)^2 \tag{39}$$

$$m_x = \frac{1}{4D_L t_o} \tag{40}$$

$$m_y = \frac{1}{4D_T t_o} \tag{41}$$

$$b_x = \ln \frac{M\sqrt{m_x m_y}}{\pi\rho} + m_y y_o^2 \tag{42}$$

$$b_y = \ln \frac{M\sqrt{m_x m_y}}{\pi\rho} + m_x x_o^2 \tag{43}$$

and where x_o and y_o are the source location, M is the mass of contaminant released per unit depth of aquifer, ρ is the density of water, D_L and D_T are longitudinal and transverse dispersion coefficients, respectively. The parameters x_m and y_o were then obtained by optimizing the correlation coefficient of the linear regression.

Another inverse analytical technique was developed by Ala and Domenico [1] to determine parameters such as the source strength and size, and the advective position of the contaminant front for the instantaneous contaminant plumes at Otis Air Force Base (AFB), Massachusetts. The contaminant plumes contained chloride, biodegradable and nonbiodegradable detergents, trichloroethene, and tetrachloroethene. To obtain the uniquely determined parameters, Ala and Domenico structured the analytical technique to solve several equations simultaneously. By assuming a uniform vertical contaminant distribution, the steady state analytical solution becomes

$$C'(x,y) = \left(\frac{C_o}{2}\right)\left\{erf\left[\frac{y+Y/2}{2(D_T x)^{1/2}}\right] - erf\left[\frac{y-Y/2}{2(D_T x)^{1/2}}\right]\right\} \tag{44}$$

where C' is the maximum concentration, D_T and Y are the unknowns. This equation holds for any point of the plume. Given two concentration values at different x-coordinates, a ratio of the maximum concentration can be obtained by

$$\frac{C'(x_1,y_1)}{C'(x_2,y_2)} = \frac{\left\{erf\left[\frac{y_1+Y/2}{2(D_T x_1)^{1/2}}\right] - erf\left[\frac{y_1-Y/2}{2(D_T x_1)^{1/2}}\right]\right\}}{\left\{erf\left[\frac{y_2+Y/2}{2(D_T x_2)^{1/2}}\right] - erf\left[\frac{y_2-Y/2}{2(D_T x_2)^{1/2}}\right]\right\}} \tag{45}$$

To solve Eq. (45), one needs to have a minimum of three wells that will give ratios of C_1'/C_2', C_1'/C_3', C_2'/C_3'. The source concentration can be found by solving Eq. (44). The results of their study indicated that the method works reasonably well with a limited amount of variation in parameters, especially the dispersivities.

Butcher and Gauthier [18] used inverse modeling to estimate the residual DNAPL mass. The flux was estimated from observed down-gradient concentrations. For the inverse model, Butcher and Gauthier used a tractable analytical approximation to the problem and developed additional simplifications to yield a form that is easily solved for the parameters of interest. From the simplified set of analytical equations, the parameters desired were solved using a least-squares estimator. The authors concluded that the methods presented in the paper could be used to provide an indication of the presence of the residual DNAPL, but not an accurate estimate of the residual volume. Improvement of the residual volume estimated can be effected using longer period of observations.

In a different field of study, namely physics, Macdonald [41] successfully applied the Nonlinear Least Squares (NLS) method to invert a 1D pure diffusion problem. He applied the method to recover a number of Dirac-delta sources with large measurement errors in the data. Alapati and Kabala [2] employed the NLS method without regularization to recover the release history of a groundwater contaminant plume from its current measured spatial distribution. The closed form solution of Eq. (2) subject to Eqs. (3–5) is [34]:

$$C(x,t) = \int_0^t C_{in}(\tau) f_r(x, t-\tau)\, d\tau \tag{46}$$

where C_{in} is the concentration release history and

$$f_r(x, t-\tau) = \frac{x}{2(\pi D(t-\tau)^3)^{1/2}} \exp\left[-\frac{(x - v(t-\tau))^2}{4D(t-\tau)}\right] \tag{47}$$

Alapati and Kabala assumed that there exist M concentration values $C_i = C(x_i, T)$ measured at time T and M measurement points x_i, with $1 \le i \le M$. Then from Eq. (46), one gets

$$F_i = C_i - \int_0^T C_{in}(\tau) f_r(x_i, T-\tau)\, d\tau = 0 \tag{48}$$

where F_i is the residual. In a compact vector representation, Eq. (48) can be written as

$$F(X) = 0 \tag{49}$$

Equation (48) leads to a system of M non-linear equations. The solutions can be obtained by using the Newton-Raphson scheme. In a number of synthetic numerical examples the solution was found to be very sensitive to noise and to the extent to which the plume is dissipated, especially for the gradual release scenario but not in the catastrophic release scenarios even in the presence of moderate measurement errors. The noise level used was 5% and 20 and 50% for moderate and high levels, respectively.

3.2.4 Direct Approaches

A different approach was proposed by Skaggs and Kabala [58]. They attempted to reconstruct the history of the plume using TR. A linear Fredholm equation of the first kind is defined by

$$y(\omega) = \int_a^b K(\lambda, \omega)\, s(\lambda)\, d\lambda \tag{50}$$

where $K(\lambda,\omega)$ and $y(\omega)$ are known and $s(\lambda)$ is the unknown to be solved. In the TR method, the ill-posed problem was replaced with a well-posed minimization problem with the function to be minimized given by

$$V(\alpha) = \left\| y(\omega) - \int_a^b K(\lambda, \omega)\, s(\lambda)\, d\lambda \right\|^2 + \alpha^2 \|Ls\|^2 \tag{51}$$

where L is the regularization operator and α is the regularization parameter that determines the relative strength of L. The most common regularization operator form is

$$\|Ls\|^2 = \int_a^b \left(\frac{d^n s}{d\lambda^n} \right)^2 d\lambda \tag{52}$$

Skaggs and Kabala studied a 1D solute transport through a saturated homogeneous medium problem with a complex contaminant release history and assumed no prior knowledge of the release function. This closed form solution is similar to Eq. (50). Due to this similarity, Eq. (46) can be utilized to estimate the concentration release history (C_T). The accuracy of the TR method depends on the regularization parameter α.

The authors pointed out that the accuracy of the regularized solution depends on finding a good value for the regularization parameter. The results showed that the solutions are insensitive to round off errors but their accuracy is affected by plume measurement errors. The authors observed that as long as the plume is not significantly dissipated, it is possible to adequately recover its history even when the data contains moderate random measurement errors. In their simulations, they used 5 and 20% noise for moderate and high levels, respectively. However, when the plume's original details are masked by dispersion, the presence of even moderate noise in the data significantly reduces the accuracy of the recovered history. This method was very effective when there existed adequate data. In a similar effort, Samarskaia [55] applied the TR with fast Fourier transforms to a groundwater contamination source reconstruction problem. Later, Liu and Ball [40] used Skaggs and Kabala's modified TR technique to study a contaminant release at Dover AFB, Delaware. They used field measured concentration profiles in low-permeability porous media that underlie a contaminated aquifer at Dover, AFB. For the two principal chemical contaminants, PCE and TCE, the Tikhonov method gave similar results as the measured data for PCE. In the case of TCE, however, it gave similar results only for the near-surface measurements. For deeper TCE concentrations the method showed anomalous behavior. Liu and Ball [40] suggested that the interpretation of the estimated results requires careful consideration in the context of other available information.

Skaggs and Kabala [59] employed the QR method for the same problem solved in their TR method. In the QR method, Skaggs and Kabala solved an equation that is close to the original equation and that is stable with a negative time step. The diffusion operator

$$\partial/\partial t - \nabla^2 \tag{53}$$

was replaced by

$$\partial/\partial t - \nabla^2 - \varepsilon\nabla^4 \tag{54}$$

A moving coordinate system was used to account for the velocity term of the ADE. A problem similar to the one presented in Skaggs and Kabala [58] is used for the QR study. The results are less accurate than that of the TR approach, but it is computationally less expensive. The authors claimed that it is much easier to incorporate heterogeneous parameters in the QR method. However, to date, heterogeneous parameters have not been incorporated either in the QR method or in the Tikhonov method by Skaggs and Kabala.

Another approach, namely a Minimum Relative Entropy (MRE) inversion was used by Woodbury et al. [70]. The MRE inversion is a method of statistical inference. Woodbury and co-workers also utilized the closed form solution, Eq. (46) and discretized it in the form

$$d_j = \sum_{n=0}^{N} g_{jn} m_n w_n \tag{55}$$

where j=1, …, M data points, m_n is the recovered release history at N time periods, w_n are the weighting functions for the numerical integration, g_{jn} corresponds to the f_r and d_j refers to the predicted $C(x,t)$. The contaminant release history is recovered by solving the linear inverse of Eq. (55).

Given prior information in terms of a lower and upper bound, a prior bias, and constraints in terms of measured data, the MRE provides exact expressions for the posterior *pdf* and expected value of the inverse problem. The plume source is also characterized by a *pdf*. The problem solved in their study is the same as Skaggs and Kabala's problem. For the noise-free data, MRE was able to reconstruct the plume evolution history indistinguishable from the true history. As for data with noise, the MRE method managed to recover the salient features of the source history. Another advantage using the MRE approach is that once the plume source history is reconstructed, future behavior of the plume can be easily predicted due to the probabilistic framework of MRE. Woodbury et al. [71] extended the MRE approach to reconstruct a 3D plume source within a 1D constant velocity field and constant dispersivity system.

Recently, Neupauer et al. [53] performed a study to compare the TR and the MRE methods. They found that both methods perform well in reconstructing a smooth source function. For an error free step function source history, the MRE performs better than the TR. On the other hand, the TR method is more robust in handling data that contain measurement errors. Neupauer and co-workers used measurement errors of 1%, 5%, and 25%.

Skaggs and Kabala [60] extended their study of TR using Monte Carlo simulation to answer the question of how far back one can use the Tikhonov procedure in recovering the release history of the plume, since the procedure always produces the recovered release curve that accurately reproduces the data. A table containing the percentage of test function recovery accuracy was produced. From the table, one can extract the information on how likely is that the recovered release history correctly resolves the details of the true release. Readers are referred to the original paper by Skaggs and Kabala [60] for the results. This approach can be applied also to solution methodologies other than TR. In addition, other factors such as transport parameter and source location uncertainty can be included in the analysis to increase the accuracy of the recovered release history.

Birchwood [17] used a Fourier based inverse technique to recover the source location and release history of the groundwater contaminant plume from breakthrough curve data obtained from a single monitoring well. The Birchwood model is capable of finding the potential release sites by imposing a causality requirement upon the release history and carefully interpreting the multiple solutions that arise thereafter. In this method, the breakthrough curve is represented as a Fourier series. The coefficients of the Fourier series are expressed as functions of the release location and the Fourier coefficients of the release history. By assuming that the Fourier series representation exists, the release history can be expressed as

$$C_{in}(t) = \frac{a_0}{2} + \sum_{n=1}^{\infty} a_n \operatorname{Re}\{e^{i\omega_n t}\} + b_n \operatorname{Im}\{\sin(e^{i\omega_n t})\} \tag{56}$$

where ω_n are the angular frequencies and a_n, b_n are coefficients to be determined.

The release location is determined by imposing a condition of causality upon the release history. Recovery of the source location and profile of the unit pulse release history with homogeneous parameters was used to demonstrate the method. Birchwood concluded that preliminary results suggest that the method is capable of quantifying the location coordinate of a pollution source to a reasonable degree of accuracy, but the results are highly sensitive to the accuracy of the inferred source location.

Recently, Atmadja [7] and Atmadja and Bagtzoglou [8, 9] proposed a method called the Marching-Jury Backward Beam equation (MJBBE) method. The Backward Beam equation (BBE) was first developed by Carasso [21] to solve mixed parabolic initial boundary problems over long time intervals. The BBE method implementation for heat equation problems was originally proposed by Buzbee and Carasso [19]. Elden [27] compared the BBE method with the Tikhonov-Phillips regularization method for parabolic equation problems. A comparison between Elden's study and the MJBBE method can be seen in Fig. 7a,b. The BBE method has also been used in the past by Buzbee and Carasso [19] to solve linear self-adjoint parabolic problems backwards in time. In addition, the method was also used by Carasso [22] in obtaining the solutions to the final value problem of Burger's equation and by Carasso et al. [24] for image restoration. For large time intervals, that is $t \to \infty$, one can assume that the solution reaches a known steady state value and one might have an approximate value of the solution at the forward terminal time, $t=T_{ft}$. Unfortunately, for this type of problems, not all unconditionally stable numerical schemes, such as the implicit and Crank-Nicolson schemes, give correct solutions. Carasso suggested that with an extra data point at $t=T_{ft}$, it is feasible to consider the use of elliptic boundary

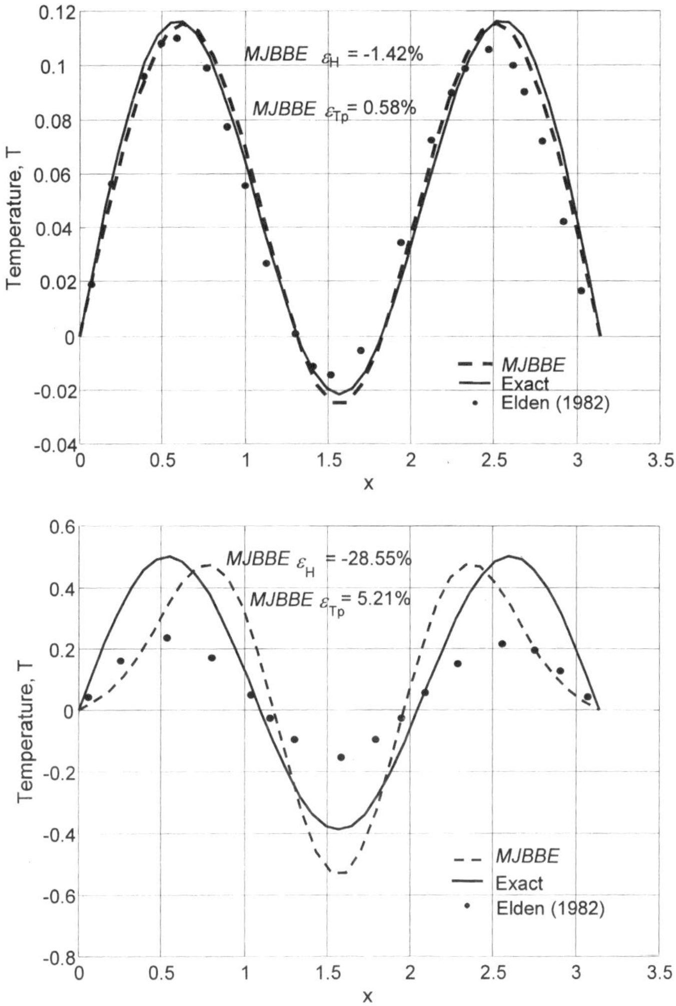

Fig. 7a, b MJBBE method results for Elden's heat equation example at: **a** t_b=0.5T_{sim}; **b** t_b=0.88T_{sim}

value techniques as a solution. His interest in the BBE method stemmed from the fact that a solution of the heat equation, f_t=f_{xx} also satisfies f_{tt}=f_{xxxx}. The method of backward beam equation was modified and enhanced to solve the 1D ADE for homogeneous and heterogeneous problems within a contamination source identification context by Atmadja and Bagtzoglou [8, 9]. Instead of solving for C directly, one needs to substitute $w=e^{kt}C$, and the auxiliary problem corresponding to Eq. (2) is

$$\frac{\partial^2 w}{\partial t^2} = \left(\frac{\partial}{\partial x} \left[D(x) \frac{\partial}{\partial x} \right] - \frac{\partial}{\partial x} [u(x)] - k \right)^2 w \quad 0 \le x \le L \quad 0 < t < T_{bi} \tag{57}$$

subject to

$$w(0,t) = w(L,t) = 0 \qquad t \geq 0 \tag{58}$$

$$w_t(0,t) = w_t(L,t) = \left[\frac{\partial}{\partial x}\left[D(x)\frac{\partial}{\partial x}\right] - \frac{\partial}{\partial x}[u(x)] - k\right] w = 0 \tag{59}$$

$$w(x,T_{bi}) = e^{kT_{bi}}C_T(x) \qquad w(x,0) = 0 \tag{60}$$

where T_{bi} is the backward initial time. k is a stabilization parameter given by

$$k = \left(\frac{1}{T_{ft}}\right)\ln\left(\frac{M}{\delta}\right) \tag{61}$$

as given by Buzbee and Carasso [19] where δ and M are bounds on the errors generated by this procedure and T_{ft} is the forward terminal time. Let such a bound be

$$\left\| f(T_{ft}) - f_2 \right\|_2 \leq \delta \tag{62}$$

for the initial data, where δ can be chosen to be the error between the measured and the exact values. Let the bound in the resulting errors in the terminal data be

$$\left\| f(0) - f_1 \right\|_2 \leq M \tag{63}$$

where M is the acceptance level for the errors in the predicted terminal values.

Atmadja and Bagtzoglou were able to cut down the computational time requirements by about three times by modifying the method to be a hybrid between marching and jury methods for the examples they presented in their work. The results for homogeneous parameters can be seen in Fig. 8. Most importantly, the MJBBE method is robust enough to handle heterogeneity. Bagtzoglou and Atmadja [13] and Atmadja [7] also compared the MJBBE method to the modified QR method for the case of heterogeneous media. Unlike Skaggs and Kabala's formulation, who used a moving coordinate system to account for the advective term of the ADE, Bagtzoglou and Atmadja applied the QR method with the advective term being explicitly handled in the ADE. The modified QR method for a 1D heterogeneous ADE is

$$\frac{\partial C}{\partial t} = \frac{\partial}{\partial x}\left[D(x)\frac{\partial C}{\partial x}\right] - \frac{\partial}{\partial x}[u(x)C] + \varepsilon\frac{\partial^4 C}{\partial x^4} \tag{64}$$

where ε is a small positive constant. The results can be seen in Fig. 9. In addition to its robustness in handling heterogeneity, the MJBBE method was able to preserve the shape and salient features of the initial data.

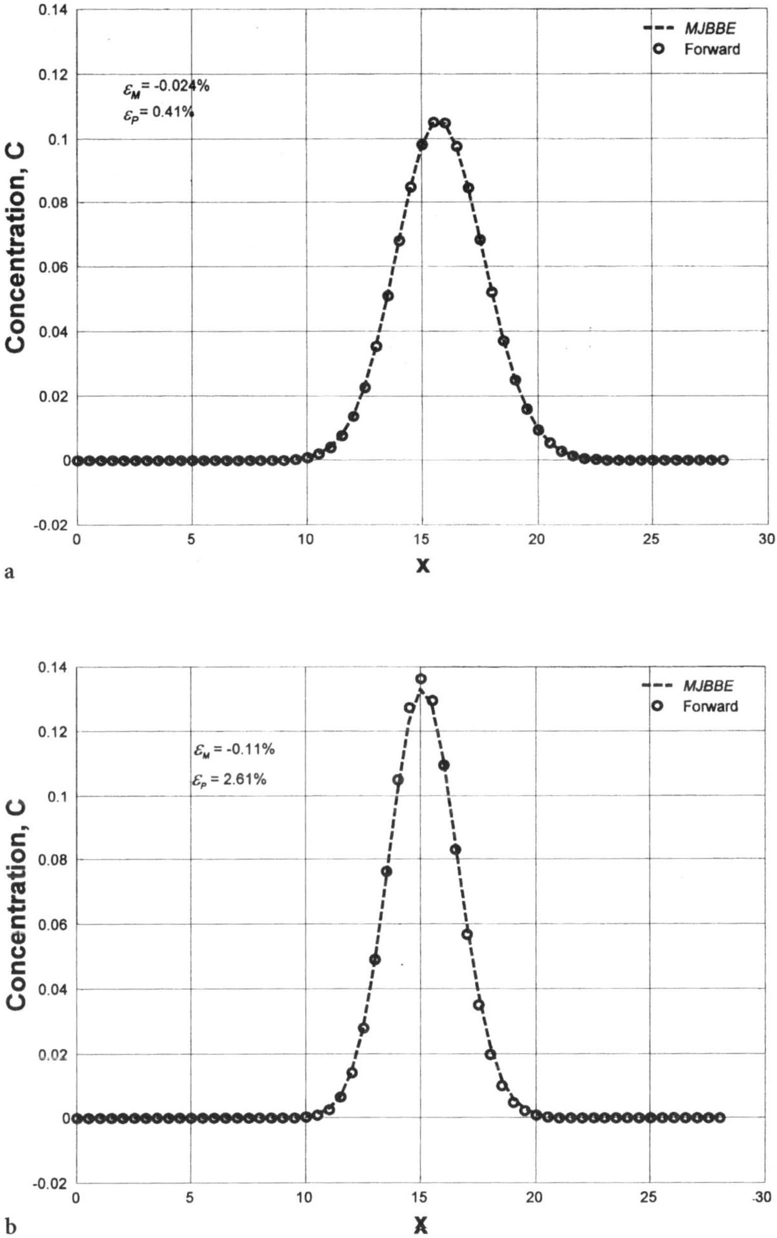

Fig. 8a, b MJBBE method results for homogeneous ADE problem: **a** at $t_b=0.2T_{sim}$; **b** at $t_b=0.9T_{sim}$

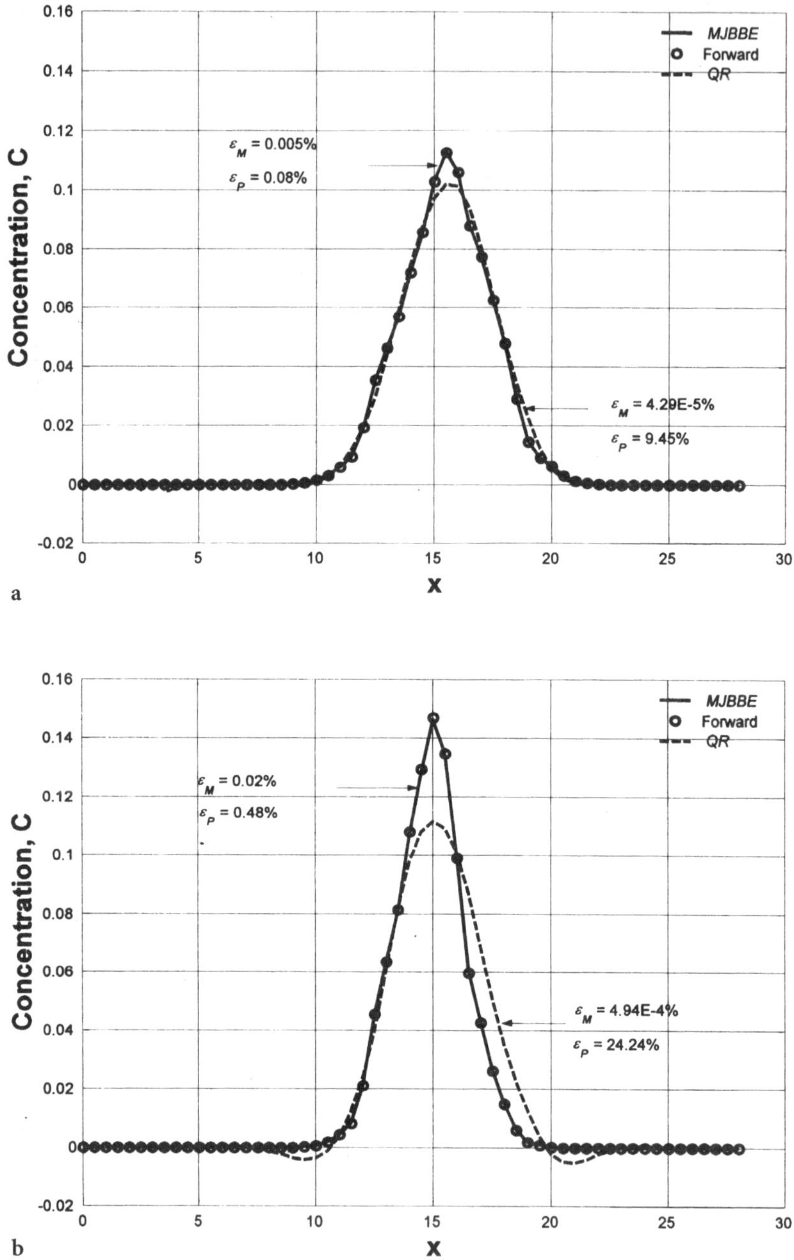

Fig. 9a, b MJBBE and QR method results for heterogeneous media at: **a** t_b=0.2T_{sim}; **b** t_b=0.9T_{sim}

4
Conclusions

Inverse modeling for source identification purposes has been around for more than three decades. Its origin was based on the needs of solving heat conduction problems backwards in time. In identifying the contaminant source location and time release history, one needs to solve the governing equation backwards in time. Due to the similarity between the heat and mass transfer processes, numerous methods used in heat conduction problems are applicable to groundwater pollution source identification. Early methods to backtrack the pollution source location were to run forward simulations and check the solutions with the measured/current spatial data observed. Solving the ADE backwards in time without relying on optimization approaches was only started in the 1990s. The methods that do not rely on optimization approaches can be grouped into three classes. One class is employing probabilistic techniques such as geostatistics to deduce the probability of the location of the sources. The second class is based on either analytical solutions or nonlinear regression approaches. The third class is using deterministic direct methods to solve the governing equations backward in time to reconstruct the release history of the contaminant plumes.

This chapter presented a review of the literature on mathematical methods used in pollution source identification.

References

1. Ala NK, Domenico PA (1992) Inverse analytical techniques applied to coincident contaminant distributions at Otis Air Force Base, Massachusetts. Ground Water 10:212–218
2. Alapati S, Kabala ZJ (2000) Recovering the release history of a groundwater contaminant via the non-linear least-squares estimation. Hydrol Proc 14:1003–1016
3. Alifanov O, Artyukhin E (1976) Regularized numerical solution of nonlinear inverse heat-conduction problem. J Engin Phys 29:934–948
4. Ames KA, Epperson JF (1997) A kernel-based method for the approximate solution of backward parabolic problems SIAM. J Numer Anal 34:1357–1390
5. Ames KA, Clark GW, Epperson JF, Oppenheimer SF (1998) A comparison of regularizations for an ill-posed problem. Math Comput 67:1451–1471
6. Aral MM, Guan J, Maslia ML (2001) Identification of contaminant source location and release history in aquifers. J Hydrol Engin 6:225–234
7. Atmadja J (2001) The marching-jury backward beam equation method and its application to backtracking non-reactive plumes in groundwater. PhD Dissertation, Columbia University, p 121
8. Atmadja J, Bagtzoglou AC (2000) Groundwater pollution source identification using the backward beam equation method in computational methods for subsurface flow and transport. AA Balkema, Rotterdam, The Netherlands, p 397–404
9. Atmadja J, Bagtzoglou AC (2001) Pollution source identification in heterogeneous porous media. Water Resour Res 37:2113–2125
10. Bagtzoglou AC (1990) Particle-grid methods with application to reacting flows and reliable solute source identification. PhD Dissertation, University of California Irvine, p 246

11. Bagtzoglou AC, Tompson AFB, Dougherty DE (1991) Probabilistic simulation for reliable solute source identification in heterogeneous porous media. Water Resources Engineering Risk Assessment, NATO ASI Series G29:189–201
12. Bagtzoglou AC, Dougherty DE, Tompson AFB (1992) Application of particle methods to reliable identification of groundwater pollution sources. Water Resour Manage 6:15–23
13. Bagtzoglou AC, Atmadja J (2001) The marching-jury backward beam equation and quasi-reversibility inverse methods for contaminant plume spatial distribution recovery. Water Resour Res 39(2):1038
14. Bayo E, Moulin H, Crisalle O, Gimenez G (1992) Well-conditioned numerical approach for the solution of the inverse heat conduction problem. Numer Heat Transfer Part B 21:79–98
15. Beck JV, Blackwell B, St Clair Jr CR (1985) Inverse heat conduction: ill-posed problems. Wiley, New York
16. Beckman FS (1960) The solution of linear equations by conjugate gradient method In: Ralston A, Wilf HS (eds) (1960) Mathematical methods for digital computers 1, chap 4. Wiley, New York, p 62–67
17. Birchwood RA (1999) Identifying the location and release characteristics of a groundwater pollution source using spectral analysis. Proceedings of the 19th Annual American Geophysical Union Hydrology Days Conference, Colorado State University, Fort Collins, Colorado, pp 37–50
18. Butcher JB, Gauthier TD (1994) Estimation of residual dense NAPL mass by inverse modeling. Ground Water 32:71–78
19. Buzbee BL, Carasso A (1973) On the numerical computation of parabolic problems for preceding times. Math Comput 27:237–266
20. Cannon JR (1966) Some numerical results for the solution of the heat equation backwards in time. Proceedings Adv Symposium Numerical Solutions of Nonlinear Differential Equations, Madison, Wisconsin, Wiley, New York, pp 21–54
21. Carraso A (1972) The backward beam equation: two A-stable schemes for parabolic problems. SIAM J Numer Anal 9:406–434
22. Carasso A (1975) The backward beam equation and the numerical computation of dissipative equations backwards in time. MRC Technical Summary Report No 1534, University of Wisconsin-Madison
23. Carasso AS (1992) Space marching difference schemes in the nonlinear inverse heat conduction problem. Inverse Problems 8:25–43
24. Carasso AS, Sanderson JG, Hyman JM (1978) Digital removal of random media image degradations by solving the diffusion equation backwards in time. SIAM J Numer Anal 15:344–367
25. Clark GW, Oppenheimer SF (1994) Quasireversibility methods for non-well-posed problems. Elec J Diff Equations 8:1–9
26. Dempster HS, Sherwood-Lollar B, Feenstra S (1997) Tracing organic contaminants in groundwater: a new methodology using compound-specific isotopic analysis. Environ Sci Technol 31:3193–3197
27. Elden L (1982) Time discretization in the backward solution of parabolic equations. II. Math Comput 39:69–84
28. Engelking P (1999) Pollution Microsoft Encarta encyclopedia 99 1993–1998. Microsoft Corporation
29. Ewing RE (1975) The approximation of certain parabolic equations backward in time by Sobolev equations. SIAM J Math Anal 6:283–294
30. Fletcher R, Reeves CM (1964) Function minimization by conjugate gradients. Brit Comput J 149–154

31. Gorelick SM, Evans BE, Remson I (1983) Identifying sources of groundwater pollution: an optimization approach. Water Resour Res 19:779–790
32. Hensel E (1991) Inverse theory and application for engineers. Prentice-Hall, Englewood Cliffs, New Jersey
33. Huang CH, Ozisik MN (1992) Inverse problem of determining unknown wall heat flux in laminar flow through a parallel plate duct. Numer Heat Trans Part A 21:55–70
34. Jury WA, Roth K (1990) Transfer functions and solute movement through soil. Birkhauser Verlag, Boston
35. Kammerer WJ, Nashed MZ (1972) On the convergence of the conjugate gradient method for singular linear operator equations. SIAM J Numer Anal 9:165–181
36. Lasdon LS, Mitter SK, Waren AD (1967) The conjugate gradient method for optimal control problem IEEE. Trans Automat Control 12:132–138
37. Lattes R, Lions JL (1969) The method of quasi-reversibility, applications to partial differential equation. Elsevier, New York
38. Li Z, Zheng K (1997) An inverse problem in a parabolic equation. J Diff Equations, Conference 01:193–199
39. Liu J, Wilson JL (1995) Modeling travel time and source location probabilities in two-dimensional heterogeneous aquifer. Proc 5th WERC Technology Development Conference, Las Cruces, New Mexico, pp 59–76
40. Liu C, Ball WP (1999) Application of inverse methods to contaminant source identification from aquitard diffusion profiles at Dover AFB. Delaware Water Resour Res 35:1975–1985
41. Macdonald JR (1995) Solution of an "impossible" diffusion-inversion problem. Comput Phys 9:546–553
42. Mahar PS, Datta B (1997) Optimal monitoring network and ground-water pollution source identification. J Water Resour Planning Manag 123:199–207
43. Mahar PS, Datta B (2000) Identification of pollution sources in transient groundwater systems. Water Resour Manag 14:209–227
44. Mahar PS, Datta B (2001) Optimal identification of ground-water pollution sources and parameter identification. J Water Resour Planning Manag 127:20–29
45. Mansuy L, Philp RP, Allen J (1997) Source identification of oil spills based on the isotopic composition of individual components in weathered oil samples. Environ Sci Technol 31:3417–3425
46. Marquardt W, Auracher H (1990) An observer-based solution of inverse heat conduction problems. Int J Heat Mass Transfer 33:1545–1562
47. Morrison RD (2000) Application of forensic techniques for age dating and source identification in environmental litigation. Environ Forensics 1:131–153
48. Morrison RD (2000) Critical review of environmental forensics techniques. I. Environ Forensics 1:157–173
49. Morrison RD (2000c) Critical review of environmental forensics techniques. II. Environ Forensics 1:173–195
50. National Research Council (1990) Groundwater models, scientific and regulatory applications. National Academy Press, Washington, DC
51. Neupauer RM, Wilson JL (1999) Adjoint method for obtaining backward-in-time location and travel time probabilities of a conservative groundwater contaminant. Water Resour Res 35:3389–3398
52. Neupauer RM, Wilson JL (2001) Adjoint-derived location and travel time probabilities for a multi-dimensional groundwater system. Water Resour Res 37:1657–1668
53. Neupauer RM, Borchers B, Wilson JL (2000) Comparison of inverse methods for reconstructing the release history of a groundwater contamination source. Water Resour Res 36:2469–2475

54. Rachdawong P, Christensen E (1997) Determination of PCB sources by a principal component method with nonnegative constraints. Environ Sci Technol 31:2686–2691
55. Samarskaia E (1995) Groundwater contamination modeling and inverse problems of source reconstruction. SAMS 18/19:143–147
56. Sidauruk P, Cheng A, Ouazar D (1998) Ground water contaminant source and transport parameter identification by correlation coefficient optimization. Ground Water 36:208–214
57. Silva Neto AJ, Ozisik MN (1993) Inverse problem of simultaneously estimating the time-wise-varying strength of two plane heat sources. J Appl Phys 73:2132–2137
58. Skaggs TH, Kabala ZJ (1994) Recovering the release history of a groundwater contaminant. Water Resour Res 30:71–79
59. Skaggs TH, Kabala ZJ (1995) Recovering the history of a groundwater contaminant plume: method of quasi-reversibility. Water Resour Res 31:2669–2673
60. Skaggs TH, Kabala ZJ (1998) Limitations in recovering the history of a groundwater contaminant plume. J Contam Hydrol 33:347–359
61. Snodgrass MF, Kitanidis PK (1997) A geostatistical approach to contaminant source identification. Water Resour Res 33:537–546
62. Stout SA, Uhler AD, Naymik TG, McCarthy KJ (1998) Environmental forensics unraveling site liability. Environ Sci Technol 32:260A–264A
63. Tikhonov AN, Arsenin VY (1977) Solutions of ill-posed problems. VH Winston, Washington, DC
64. US Environmental Protection Agency (1998) Report Brochure: National Water Quality Inventory: 1996 Report to Congress, Background Section [online]. Office of Water, Washington DC Available: http://www.epa.gov/OW/resources/brochure/broch2.html (28 March 2000)
65. US Environmental Protection Agency (1998) National Water Quality Inventory: 1996 Report to Congress, Ground Water Chapters [online]. Office of Water, Washington DC Available: http://www.epa.gov/OW/resources/9698/chap6.html (28 March 2000)
66. US Environmental Protection Agency (1998c) This is Superfund. Available: http://www.epa.gov/superfund/whatissf/sfguide.htm (20 March 2000)
67. US Environmental Protection Agency (1999) Safe Drinking Water Act, Section 1429. Groundwater Report to Congress [online]. Office of Water, Washington DC EPA-816-R-99-016. Available: http://www.epa.gov/safewater/gwr/finalgw.pdf (28 March 2000)
68. Wagner BJ (1992) Simultaneously parameter estimation and contaminant source characterization for coupled groundwater flow and contaminant transport modeling. J Hydrol 135:275–303
69. Wilson JL, Liu J (1994) Backward tracking to find the source of pollution. Waste Manag Risk Remediation 1:181–199
70. Woodbury AD, Ulrych TJ (1996) Minimum relative entropy inversion: theory and application to recovering the release history of groundwater contaminant. Water Resour Res 32:2671–2681
71. Woodbury A, Sudicky E, Ulrych TJ, Ludwig R (1998) Three-dimensional plume source reconstruction using minimum relative entropy inversion. J Contam Hydrol 32:131–158

Handb Environ Chem Vol. 5, Part F, Vol. 3 (2005): 97–132
DOI 10.1007/b11433

Nonaqueous Phase Liquid Pool Dissolution in Subsurface Formations

Constantinos V. Chrysikopoulos (✉)

Department of Civil Engineering, University of Patras, Rio 26500, Greece
gios@upatras.gr

1 Introduction . . . 100

2 Mass Flux at the NAPL-Water Interface . . . 101

3 Analytical Models for NAPL Pool Dissolution and Contaminant Transport in Porous Media . . . 104
3.1 Rectangular Pool . . . 105
3.2 Elliptic/Circular Pool . . . 106
3.3 Model Simulations . . . 107

4 NAPL Pool Dissolution in Heterogeneous Formations . . . 108
4.1 Two-Dimensional Numerical Model . . . 108
4.2 Numerical Model Simulations . . . 111

5 NAPL Pool Dissolution in the Presence of Dissolved Humic Substances . . . 113
5.1 Two-Dimensional Solute-Humic Cotransport Numerical Model . . . 113
5.1.1 Solute Transport and Effective Mass Transfer . . . 113
5.1.2 Transport of Dissolved Humic Substances . . . 116
5.1.3 Transport of Solute-Humic Substances . . . 117
5.2 Numerical Model Simulations . . . 118

6 Mass Transfer Correlations . . . 119
6.1 Local Mass Transfer Correlations . . . 120
6.1.1 Rectangular Pools . . . 120
6.1.2 Elliptic/Circular Pools . . . 122
6.2 Average Mass Transfer Correlations . . . 123
6.2.1 Rectangular Pools . . . 123
6.2.2 Elliptic/Circular Pools . . . 124

7 Experimental Studies of NAPL Pool Dissolution in Porous Media . . . 125
7.1 TCE Dissolution Data . . . 126
7.2 Experimental Mass Transfer Correlation . . . 127

8 Summary . . . 128

References . . . 129

Abstract The objective of this chapter is to present some recent developments on nonaqueous phase liquid (NAPL) pool dissolution in water saturated subsurface formations. Closed form analytical solutions for transient contaminant transport resulting from the dissolution of a single component NAPL pool in three-dimensional, homogeneous porous media are presented for various shapes of source geometries. The effect of aquifer anisotropy and heterogeneity as well as the presence of dissolved humic substances on mass transfer from a NAPL pool is discussed. Furthermore, correlations, based on numerical simulations as well as available experimental data, describing the rate of interface mass transfer from single component NAPL pools in saturated subsurface formations are presented.

Keywords NAPL pool dissolution · Mass transfer correlations · Cotransport · Mathematical modeling

List of Symbols and Abbreviations

a	Major semi-axis of elliptic pool, L
b	Minor semi-axis of elliptic pool, L
C	Liquid phase solute concentration (solute mass/liquid volume), M/L^3
C_a	Apparent aqueous phase solute concentration in the presence of doc, M/L^3
C_b	Constant background aqueous phase solute concentration, M/L^3
C_s	Aqueous saturation concentration (solubility), M/L^3
C_Y	Covariance function of Y
C^*	Sorbed solute concentration onto the solid matrix (solute mass/solids mass)
C^*_{doc}	Solute concentration sorbed onto doc (solute mass/liquid volume), M/L^3
C^{**}_{doc}	Solute humic concentration sorbed onto the solid matrix (solute mass/solids mass)
$\mathcal{D}$	Molecular diffusion coefficient, L^2/t
$\mathcal{D}_e$	Effective molecular diffusion coefficient, equal to $\mathcal{D}/\tau^*$, L^2/t
D_x	Longitudinal hydrodynamic dispersion coefficients, L^2/t
D_y	Lateral hydrodynamic dispersion coefficients, L^2/t
D_z	Vertical hydrodynamic dispersion coefficient, L^2/t
$erf[\eta]$	Error function, equal to $(2/\pi^{1/2})\int_0^{\eta} e^{-\zeta^2} d\zeta$
f	Arbitrary function
h	Total head potential, L
H	Concentration of humic substances in the aqueous phase expressed as doc, M/L^3
H_0	Source concentration of humic substances expressed as doc, M/L^3
k	Local mass transfer coefficient, defined in Eq. (3), L/t
$\hat{k}$	Time invariant local mass transfer coefficient, defined in Eq. (4), L/t
$\bar{k}$	Time dependent average mass transfer coefficient, defined in Eq. (5), L/t
k^*	Time invariant average mass transfer coefficient, defined in Eq. (6), L/t
k_e	Effective local mass transfer coefficient, defined in Eq. (46), L/t
$\bar{k}_e$	average effective mass transfer coefficient, defined in Eq. (47), L/t
K	Hydraulic conductivity, L/t
K_d	Solute partition coefficient between the solid matrix and the aqueous phase, L^3/M
K_{doc}	Solute partition coefficient between the doc and aqueous phase, L^3/M
K_h	Humic substance partition coefficient between solid matrix and aqueous phase, L^3/M
ℓ_c	Characteristic length, L
ℓ_x	Pool dimension in x direction, L
ℓ_{x_0}	x Cartesian coordinate of the origin of a rectangular pool or the center of an elliptic/circular pool, L

ℓ_y	Pool dimension in y direction, L
ℓ_{y_o}	y Cartesian coordinate of the origin of a rectangular pool or the center of an elliptic/circular pool, L
L_h	Humic substance source height, L
m_1, m_2	Defined in Eqs. (22) and (23), respectively
n_1, n_2	Defined in Eqs. (24) and (25), respectively
N	Number of different realizations of the hydraulic conductivity field
Pe_x	Local Peclet number in the longitudinal direction
Pe_x^*	Average Peclet number in the longitudinal direction
Pe_y	Local Peclet number in the transverse direction
Pe_y^*	Average Peclet number in the transverse direction
r	Radius of circular pool, L
r_x	Separation distance in the longitudinal direction of two K measurements, L
r_z	Separation distance in the vertical direction of two K measurements, L
r	Vector with magnitude equal to the separation distance of two K measurements
R	Dimensionless retardation factor
R_s	Dimensionless retardation factor of a solute in the presence of humics, defined in Eq. (45)
R_h	Dimensionless retardation factor of the humic substances, defined in Eq. (50)
R_{sh}	Dimensionless retardation factor of the solute humic substances, defined in Eq. (56)
$R_{(e)}$	Domain defined by an elliptic NAPL-water interfacial area
$R_{(r)}$	Domain defined by a rectangular NAPL-water interfacial area
Sh	Local Sherwood number
Sh^*	Average Sherwood number
t	Time, t
U_x	Interstitial fluid velocity in the longitudinal direction, L/t
U_z	Interstitial fluid velocity in the vertical direction, L/t
U	Interstitial velocity vector
v	Defined in Eq. (26)
x	Spatial coordinate, L
y	Spatial coordinate, L
Y	Logtransformed hydraulic conductivity, equal to $ln\ K$
$\bar{Y}$	Mean logtransformed hydraulic conductivity
z	Spatial coordinate, L
α_L	Longitudinal dispersivity, L
α_T	Transverse dispersivity, L
$\beta_1, \beta_2, \beta_3$	Empirical coefficients
$\gamma_1, \gamma_2, \gamma_3$	Empirical coefficients
ε	Eccentricity, defined in Eq. (27)
ζ_x	Correlation length scale in the longitudinal direction, L
ζ_z	Correlation length scale in the vertical direction, L
θ	Porosity (liquid volume/aquifer volume), L^3/L^3
κ_1, κ_2	Defined in Eqs. (16) and (17), respectively
λ	First-order decay coefficient, t^{-1}
μ	Dummy integration variable
μ_1, μ_2	Defined in Eqs. (20) and (21), respectively
ξ_1, ξ_2	Defined in Eqs. (18) and (19), respectively
ρ_b	Bulk density of the solid matrix, M/L^3
σ_Y^2	Variance of ln$K$
τ	Dummy integration variable

τ^*	Tortuosity coefficient (≥1)
ADI	Alternating direction implicit
doc	Dissolved organic carbon
DNAPL	Dense nonaqueous phase liquid
NAPL	Nonaqueous phase liquid
PCE	Perchloroethylene
TCE	Trichloroethylene
1,1,2-TCA	1,1,2-Trichloroethane

1 Introduction

The contamination of natural subsurface systems by nonaqueous phase liquids (NAPLs), has captured the attention of many environmental engineers and scientists. Most of the NAPLs are organic solvents and petroleum hydrocarbons originating from leaking underground storage tanks, ruptured pipelines, surface spills, hazardous waste landfills, disposal sites, and leachates from recycled wastes. When an NAPL spill infiltrates the subsurface environment through the vadose zone, a portion of it may be trapped and immobilized within the unsaturated porous formation in the form of blobs or ganglia, which are no longer connected to the main body of the nonaqueous phase liquid. Upon reaching the water table, dense nonaqueous phase liquids (DNAPLs) with densities heavier than that of water (sinkers, e.g., organic leachates from recycled hazardous solid wastes), given that the pressure head at the capillary fringe is sufficiently large, continue to migrate downward leaving behind trapped ganglia until they encounter an impermeable layer, where a flat source zone or pool starts to form [1–3]. On the other hand, NAPLs with densities lower than that of water (floaters, e.g., petroleum products) as soon as they approach the saturated region, spread laterally and float on the water table in the form of a pool [4].

As groundwater flows past trapped ganglia or NAPL pools, a fraction of the NAPL dissolves in the aqueous phase and a plume of dissolved hydrocarbons is created as illustrated in Fig. 1. The dissolved NAPL concentrations may reach saturation or near saturation levels [5]. It should be noted that the dissolution of NAPL pools in porous media is fundamentally different than that of residual blobs. NAPL pools have limited contact areas with respect to groundwater. If the same amount of a NAPL is present as ganglia and as pool, ganglia dissolution is expected to proceed at a faster rate, because of the larger surface area available for interphase mass transfer [6]. Therefore, NAPL pools often lead to longlasting sources of groundwater contamination.

The concentration of a dissolved NAPL in groundwater is governed mainly by interface mass-transfer processes that often are slow and rate-limited [7, 8]. There is a relatively large body of available literature on the migration of NAPLs and dissolution of residual blobs [6, 9–22], and pools [5, 23–34]. Furthermore, empirical correlations useful for convenient estimation of NAPL dissolution

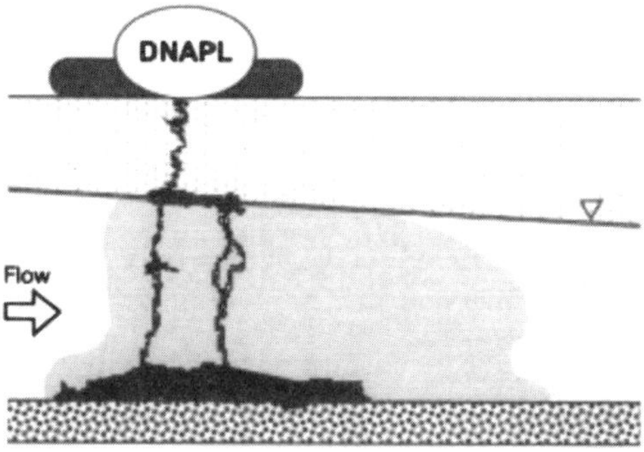

Fig. 1 Schematic illustration of slightly soluble in water dense nonaqueous phase liquid migration in the subsurface and plume formation of dissolved hydrocarbons

rates, expressed in terms of non dimensional parameters, for well defined residual blobs are available in the literature [11, 15, 35]. These correlations relate the Reynolds number and NAPL volumetric fraction to the dimensionless mass transfer rate coefficient (Sherwood number), and differ mainly in the number and type of dimensionless system properties accounted for. The existing correlations indicate that dissolution rates for residual NAPLs in one-dimensional soil columns are highly dependent on blob shape and size, interstitial velocity, and/or length of soil zone exposed to residual NAPLs [8, 15].

Mathematical models for mass transfer at the NAPL-water interface often adopt the assumption that thermodynamic equilibrium is instantaneously approached when mass transfer rates at the NAPL-water interface are much faster than the advective-dispersive transport of the dissolved NAPLs away from the interface [28, 36]. Therefore, the solubility concentration is often employed as an appropriate concentration boundary condition specified at the interface. Several experimental column and field studies at typical groundwater velocities in homogeneous porous media justified the above equilibrium assumption for residual NAPL dissolution [9, 37–39].

2
Mass Flux at the NAPL-Water Interface

As a NAPL pool dissolves into the interstitial fluid of a water saturated porous formation, a concentration boundary layer is assumed to be developed above the NAPL-water interface. Assuming that the thickness of the pool is insignificant relative to the thickness of the aquifer, the mass transfer from the NAPL-water interface into the aqueous interstitial fluid within a three-dimensional, saturated porous formation is described by the following relationship [40]:

$$-\mathcal{D}_e \left. \frac{\partial C(t,x,y,z)}{\partial z} \right|_{z \to 0} = k(t,x,y)[C_s - C(t,x,y,\infty)] \tag{1}$$

where $C(t, x, y, z)$ is the aqueous phase solute concentration; x, y, z are the spatial coordinates in the longitudinal, transverse and vertical (perpendicular to the interface) directions, respectively; t is time; $D_e = D/\tau^*$ is the effective molecular diffusion coefficient (where D is the molecular diffusion coefficient, and $\tau^* \geq 1$ is the tortuosity coefficient); $k(t, x, y)$ is the local mass transfer coefficient dependent on time and location at the NAPL-water interface; and C_s is the aqueous saturation (solubility) concentration of the NAPL at the interface. Conventionally, any location above the concentration boundary layer is considered as $z \rightarrow \infty$. For the case where the background concentration is constant with respect to time and space, for notational convenience, $C(t, x, y, \infty)$ is replaced by C_b, the constant background aqueous phase concentration. It should be noted that in all of the cases considered in this chapter the free stream concentration is assumed to be zero ($C_b = 0$). The left-hand side term in Eq. (1) represents a diffusive flux given by Fick's law, whereas the right-hand side represents a convective mass transfer flux. The mass transfer relationship given by Eq. (1) implies that the dissolution at the NAPL-water interface is fast. Therefore, the aqueous phase solute concentration is limited only by mass transfer. The concentration along the interface is assumed constant and equal to the saturation concentration,

$$C(t, x, y, 0) = C_s \tag{2}$$

The development of a concentration boundary layer at steady-state uniform flow conditions for two different constant background liquid phase concentrations is shown in Fig. 2. The NAPL-water interface is indicated by the x-axis, whereas the concentration within the boundary layer is shown with arrows. Traditionally, arrows are reserved for the representation of vectors; however, arrows are employed in Fig. 2 for convenient sketching of the magnitude of the scalar concentration. The first case (Fig. 2a) illustrates the situation where the background liquid phase concentration is zero and the concentration within the boundary layer decreases from saturation concentration at the NAPL-water interface to zero concentration in the bulk interstitial liquid. The second case (Fig. 2b) represents the situation where the background concentration is constant, but nonzero, and the concentration within the boundary layer varies from the solubility limit at the interface to a finite constant concentration in the bulk interstitial fluid. The thickness of the boundary layer depends mainly on the hydrodynamics and dispersion properties of the system, and is defined as the value of z for which $[C(t, x, y, 0) - C(t, x, y, z)]/[C(t, x, y, 0) - C(t, x, y, \infty)] = 0.99$ [41]. Examination of Fig. 2 indicates that a thin boundary layer corresponds to steeper concentration gradients. It should be noted, however, that the greater the concentration gradient the greater the mass transfer of the dissolved component. Therefore, the local mass transfer coefficient decreases with distance from the front end of the NAPL pool and has a maximum value at the leading or upstream edge. The local mass transfer coefficient is not only spatially but temporally dependent during the initial period of formation of the boundary layer. At steady-state physicochemical and hydrodynamic conditions the local mass transfer coefficient is independent of time.

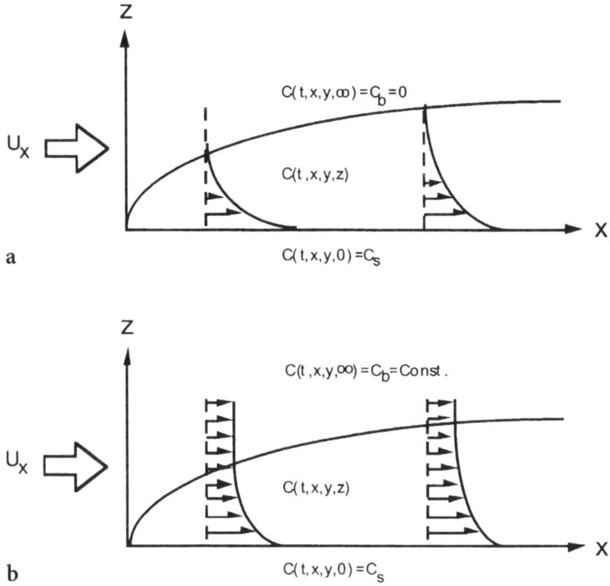

Fig. 2a, b Development of a concentration boundary layer along the NAPL-water interface in uniform interstitial flow for: **a** zero bulk aqueous phase (background) concentration; **b** constant nonzero background concentration (adopted from Chrysikopoulos and Lee [30])

In view of Eq. (1), the time and space dependent local mass transfer coefficient is given by

$$k(t, x, y) = -\frac{\mathcal{D}_e}{C_s} \frac{\partial C(t, x, y, z)}{\partial z}\bigg|_{z \to 0} \tag{3}$$

At the relatively rare case where steady state physicochemical and hydrodynamic conditions exist, the local mass transfer coefficient is independent of time,

$$\hat{k}(x, y) = -\frac{\mathcal{D}_e}{C_s} \frac{\partial C(x, y, z)}{\partial z}\bigg|_{z \to 0} \tag{4}$$

It should be noted that the local mass transfer coefficient can only be obtained experimentally and is case specific. An analytical relationship for the local mass transfer rate coefficient can be obtained if a mathematical expression describing the gradient of the dissolved concentration at the NAPL-water interface is known. Unfortunately, the local mass transfer coefficient usually is not an easy parameter to determine with precision. Thus, in mathematical modeling of contaminant transport originating from NAPL pool dissolution, $k(t, x, y)$ is often replaced by the average mass transfer coefficient, $\bar{k}(t)$, applicable to the entire pool, expressed as [41]

$$\bar{k}(t) = \frac{1}{A} \int_A k(t, x, y)\, \mathrm{d}^2\mathbf{A} \tag{5}$$

where A is the surface area of the NAPL pool, and $d^2\mathbf{A}$ is a differential surface area. Furthermore, at steady state conditions $k(t, x, y)$ is replaced by the corresponding time invariant, average mass transfer coefficient, k^*, applicable to the entire pool, expressed as [30]

$$k^* = \frac{1}{A} \int_A \hat{k}(x, y)\, d^2\mathbf{A} \tag{6}$$

3
Analytical Models for NAPL Pool Dissolution and Contaminant Transport in Porous Media

The transient contaminant transport from a dissolving DNAPL pool in a water saturated, three-dimensional, homogeneous porous medium under steady-state uniform flow, assuming that the dissolved organic sorption is linear and instantaneous, is governed by the following partial differential equation:

$$R\frac{\partial C(t,x,y,z)}{\partial t} = D_x \frac{\partial^2 C(t,x,y,z)}{\partial x^2} + D_y \frac{\partial^2 C(t,x,y,z)}{\partial y^2} + D_z \frac{\partial^2 C(t,x,y,z)}{\partial z^2} - U_x \frac{\partial C(t,x,y,z)}{\partial x} - \lambda R C(t,x,y,z) \tag{7}$$

where U_x is the average unidirectional interstitial fluid velocity; D_x, D_y, D_z are the longitudinal, transverse, and vertical hydrodynamic dispersion coefficients, respectively, defined as [42]

$$D_x = \alpha_L U_x + \mathcal{D}_e \tag{8a}$$

$$D_y = \alpha_T U_x + \mathcal{D}_e \tag{8b}$$

$$D_z = \alpha_T U_x + \mathcal{D}_e \tag{8c}$$

where α_L and α_T are the longitudinal and transverse dispersivities, respectively; R is the dimensionless retardation factor accounting for the effects of sorption, representing the ratio of the average interstitial fluid velocity to the propagation velocity of the solute [43], and for linear, instantaneous sorption is given by

$$R = 1 + \frac{\rho_b K_d}{\theta} \tag{9}$$

(where K_d is the partition or distribution coefficient; ρ_b is the bulk density of the solid matrix; and θ is the porosity of the porous medium); and λ is a first-order decay constant. It should be noted that the decay term λRC in the governing Eq. (7) indicates that the total concentration (aqueous plus sorbed solute mass) disappears due to possible decay or biological/chemical degradation.

3.1 Rectangular Pool

For a DNAPL pool with rectangular geometry as shown in Fig. 3a, assuming that NAPL pool dissolution is described by Eq. (1), the appropriate initial and boundary conditions are

$$C(0, x, y, z) = 0 \tag{10}$$

$$C(t, \pm\infty, y, z) = 0 \tag{11}$$

$$C(t, x \pm \infty, z) = 0 \tag{12}$$

$$\mathcal{D}_e \frac{\partial C(t, x, y, z)}{\partial z} = \begin{cases} -k(t, x, y)\, C_s & \ell_{x_o} < x < \ell_{x_o} + \ell_x, \\ & \ell_{y_o} < y < \ell_{y_o} + \ell_y, \\ 0 & \end{cases} \tag{13}$$

Otherwise,

$$C(t, x, y, \infty) = 0 \tag{14}$$

where ℓ_{x_o}, ℓ_{y_o} indicate the x, y Cartesian coordinates of the pool origin, respectively, and ℓ_x, ℓ_y are the pool dimensions in x, y directions, respectively. The analytical solution to the governing partial differential Eq. (7) subject to conditions given by Eqs. (10)–(14) for the special case where $k(t, x, y)=k^*$ has been derived by Chrysikopoulos [40] as follows:

$$C(t, x, y, z) = \frac{C_s k^*}{4\mathcal{D}_e} \int_0^t \left(\frac{D_z}{R\pi\tau} \right)^{1/2} \exp\left[-\lambda\tau - \frac{Rz^2}{4D_z\tau} \right] (\mathrm{erf}[\kappa_1] - \mathrm{erf}[\kappa_2])\, (\mathrm{erf}[\xi_1] - \mathrm{erf}[\xi_2])\, d\tau \tag{15}$$

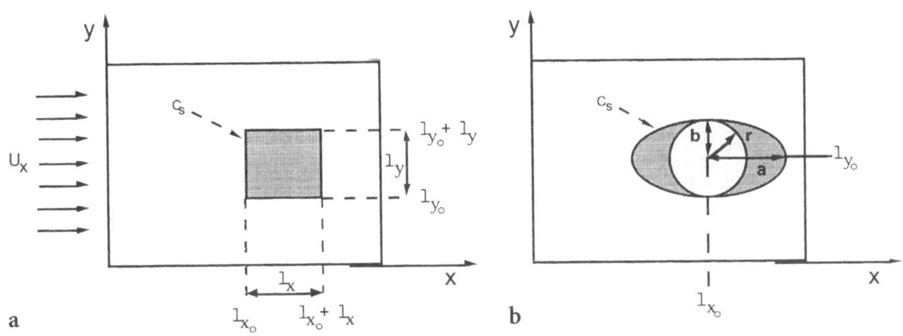

Fig. 3a, b Plan view of a denser than water: **a** rectangular NAPL pool with dimensions $\ell_x \times \ell_y$; **b** elliptic NAPL pool with origin at $x = \ell_{x_o}, y = \ell_{y_o}$ having major semi-axis a and minor semi-axis b. For the special case where $a=b=r$ the elliptic pool becomes a circular pool with radius r

where

$$\kappa_1 = \left(x - \ell_{x_o} - \frac{U_x\tau}{R}\right)\left(\frac{R}{4D_x\tau}\right)^{1/2} \tag{16}$$

$$\kappa_2 = \left(x - \ell_{x_o} - \ell_x - \frac{U_x\tau}{R}\right)\left(\frac{R}{4D_x\tau}\right)^{1/2} \tag{17}$$

$$\xi_1 = (y - \ell_{y_o})\left(\frac{R}{4D_y\tau}\right)^{1/2} \tag{18}$$

$$\xi_2 = (y - \ell_{y_o} - \ell_y)\left(\frac{R}{4D_y\tau}\right)^{1/2} \tag{19}$$

$erf[\ ]$ is the error funktion defined as $erf[\eta] = (2/\pi^{1/2})\int_0^{\eta} e^{-\zeta^2} d\zeta$; and τ is a dummy integration variable. Note that in view of Eq. (1) it is evident that the analytical solution given by Eq. (15) can also be employed for cases of finite but constant free stream concentration (C_b=const.$\neq$0) by replacing C_s by C_s–C_b.

3.2 Elliptic/Circular Pool

For a DNAPL pool with elliptic geometry as shown in Fig. 3b, assuming that NAPL pool dissolution is described by Eq. (1), the appropriate initial and boundary conditions are given by Eqs. (10)–(12), (14) and

$$\mathcal{D}_e \frac{\partial C(t,x,y,0)}{\partial z} = \begin{cases} -k(t,x,y)\, C_s \sim \dfrac{(x-\ell_{x_o})^2}{a^2} + \dfrac{(y-\ell_{y_o})^2}{b^2} \leq 1 \\ 0 \qquad \text{otherwise} \end{cases} \tag{20}$$

where a and b are the major and minor semi-axes of the elliptic pool, respectively; and ℓ_{x_o} and ℓ_{y_o} indicate the x and y Cartesian coordinates of the pool origin, respectively.

The analytical solution to the governing partial differential Eq. (7) subject to conditions given by Eqs. (10)–(12), (14), and (20) for the special case where $k(t, x, y)$=k^* has been derived by Chrysikopoulos [40] as follows:

$$C(t,x,y,z) = \frac{C_s k^*}{4\mathcal{D}_e}\int_0^t\int_{m_1}^{m_2}\left(\frac{D_z}{R\tau}\right)^{1/2} \exp\left[-\lambda\tau - \frac{Rz^2}{4D_z\tau}\right] \times \exp[-\mu^2]\,(\text{erf}[-n_2] - \text{erf}[n_1])\, d\mu\, d\tau, \tag{21}$$

where

$$m_1 = (y - \ell_{y_o} + b)\left(\frac{R}{4D_y\tau}\right)^{1/2} \tag{22}$$

$$m_2 = (y - \ell_{y_o} - b)\left(\frac{R}{4D_y\tau}\right)^{1/2} \tag{23}$$

$$n_1 = \left[x - \frac{U_x\tau}{R} - \ell_{x_o} + \left\{\left[1 - \frac{(v - \ell_{y_o})^2}{b^2}\right]a^2\right\}^{1/2}\right]\left(\frac{R}{4D_x\tau}\right)^{1/2} \tag{24}$$

$$n_2 = \left[x - \frac{U_x\tau}{R} - \ell_{x_o} - \left\{\left[1 - \frac{(v - \ell_{y_o})^2}{b^2}\right]a^2\right\}^{1/2}\right]\left(\frac{R}{4D_x\tau}\right)^{1/2} \tag{25}$$

$$v = y - \mu\left(\frac{4D_y\tau}{R}\right)^{1/2} \tag{26}$$

μ and τ are dummy integration variables. Note that in view of Eq. (1) it is evident that the analytical solution given by Eq. (21) can also be employed for cases of finite but constant free stream concentration (C_b=const.$\neq$0) by replacing C_s by C_s-C_b.

A circular pool with radius r, as shown in Fig. 3b, is equivalent to an elliptic pool when $a=b=r$. Because the circular pool geometry is a special case of an elliptic pool, the appropriate solution to the case of a circular pool can be obtained directly from Eq. (21) by substituting $a=b=r$.

3.3 Model Simulations

The analytical solution for a rectangular pool, Eq. (15), is employed to simulate dissolved concentrations originating from the NAPL pool along the rectangular-pool centerline in the downstream direction at z=0.04 m from three NAPL pool sources of equal surface area with dimensions $\ell_x \times \ell_y$ of 1×9 m^2, 3×3 m^2, and 9×1 m^2, respectively. The results are presented in Fig. 4a and clearly indicate that the source structure and orientation control the concentration level of the dissolved NAPL. The longer the pool in the direction of flow, the higher the dissolved peak concentration in the interstitial fluid. However, the wider the pool perpendicular to the direction of flow, the larger the immediate downstream area covered by the dissolved contaminant.

The analytical solution for an elliptic pool, Eq. (21), is employed to investigate the effect of eccentricity, ε, of an elliptic NAPL pool on dissolved contaminant concentration. The model simulations are presented in Fig. 4b. It should be noted that the eccentricity of an elliptic pool is defined as

$$\varepsilon = \left[1 - \left(\frac{b}{a}\right)^2\right]^{1/2} \tag{27}$$

When ε=0, then $a=b$ and the elliptic pool becomes circular. The closer ε is to one, the more elongated the elliptic pool. Figure 4b shows that the dissolved conta-

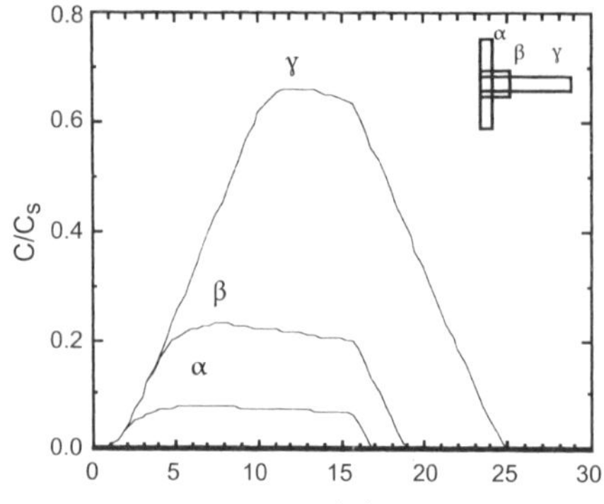

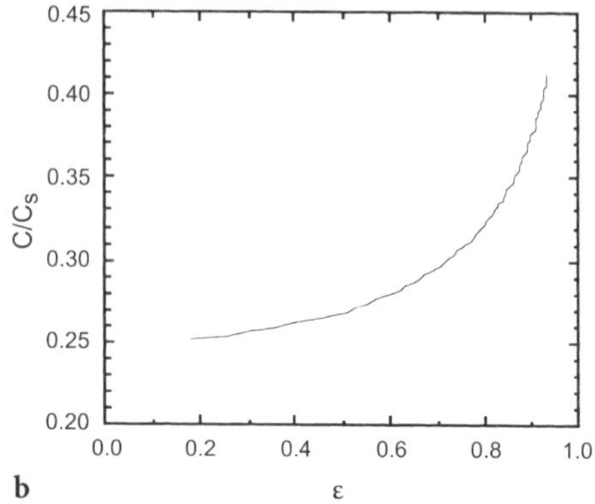

Fig. 4a, b Distribution of normalized aqueous phase NAPL concentration as a function of downstream distance for pools of equal surface area and dimensions $\ell_x \times \ell_y$ (α) 1×9 m^2, (β) 3×3 m^2, and (γ) 9×1 m^2, (here ℓ_{x_o}=0 m, ℓ_{y_o}=0, 3, 4 m for cases α, β, and γ, respectively, y=4.5 m, z=0.04 m); (b) Normalized aqueous phase NAPL concentration predicted at a location with coordinates (x, y, z)=(10, 5, 0.04 m) as a function of eccentricity of an elliptic pool (here $\ell_{x_o}=\ell_{y_o}$=5 m)

minant concentration at (x, y, z)=(10, 5, 0.04 m) within a three-dimensional homogeneous porous formation is increasing with increasing ε. Certainly, the result of Fig. 4b is in perfect agreement with that of Fig. 4a.

4 NAPL Pool Dissolution in Heterogeneous Formations

The effects of aquifer anisotropy and heterogeneity on NAPL pool dissolution and associated average mass transfer coefficient have been examined by Vogler and Chrysikopoulos [44]. A two-dimensional numerical model was developed to determine the effect of aquifer anisotropy on the average mass transfer coefficient of a 1,1,2-trichloroethane (1,1,2-TCA) DNAPL pool formed on bedrock in a statistically anisotropic confined aquifer. Statistical anisotropy in the aquifer was introduced by representing the spatially variable hydraulic conductivity as a log-normally distributed random field described by an anisotropic exponential covariance function.

4.1 Two-Dimensional Numerical Model

The transport of a sorbing contaminant in two-dimensional, heterogeneous, saturated porous media, resulting from the dissolution of a single component NAPL pool, as illustrated in Fig. 5, is described by the following partial differential equation [44]:

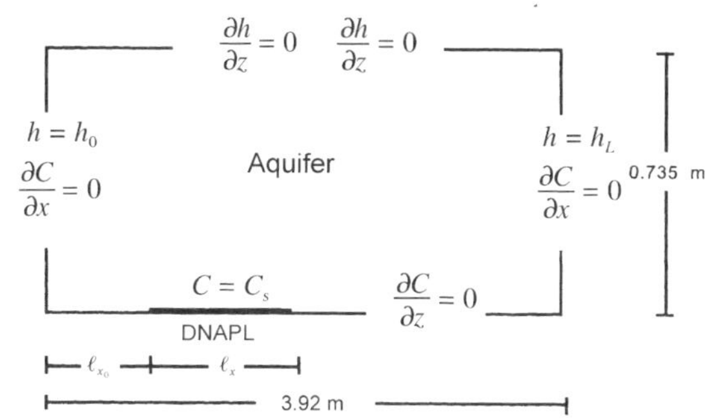

Fig. 5 Schematic illustration of the physical system examined, showing the DNAPL pool location and the appropriate boundary conditions employed in the numerical model

$$R\frac{\partial C(t,x,z)}{\partial t}=\frac{\partial}{\partial x}\left[D_x(x,z)\frac{\partial C(t,x,z)}{\partial x}\right]+\frac{\partial}{\partial z}\left[D_z(x,z)\frac{\partial C(t,x,z)}{\partial z}\right]$$
$$-\frac{\partial}{\partial x}[U_x(x,z)C(t,x,z)]-\frac{\partial}{\partial z}[U_z(x,z)C(t,x,z)], \quad (28)$$

where U_z is the vertical interstitial fluid velocity. It was assumed that the NAPL phase is immobile, and aquifer anisotropy is governed by the variability in the hydraulic conductivity field. The two-dimensional hydraulic conductivity field was generated stochastically with the computer program SPRT2D [45]. Furthermore, it was assumed that the hydraulic conductivity follows a log-normal distribution and the log-transformed hydraulic conductivity, Y, varies spatially according to the following anisotropic exponential covariance function [27, 46–48]:

$$C_Y(\mathbf{r})=\sigma_Y^2\exp\left[-\left(\frac{r_x^2}{\zeta_x^2}+\frac{r_x^2}{\zeta_z^2}\right)^{1/2}\right] \quad (29)$$

where $\mathbf{r}=(r_x, r_z)^T$ is a two-dimensional vector whose magnitude is the separation distance of two hydraulic conductivity measurements; ζ_x and ζ_z are the correlation length scales in the x and z directions, respectively; and σ_Y^2 is the variance of the log transformed hydraulic conductivity defined as

$$Y=\ln K \quad (30)$$

where K is the hydraulic conductivity. It should be noted that the stochastic realizations of $K(x,z)$ are locally independent of direction, and that anisotropy is introduced globally by the overall spatial variability of the hydraulic conductivity field [49].

For each hydraulic conductivity field generated, the associated variable hydraulic head field and groundwater velocity field were determined. Each variable hydraulic head field was evaluated numerically by solving the following steady state two-dimensional groundwater flow equation for a heterogeneous confined aquifer [50]:

$$\frac{\partial}{\partial x}\left[K(x,z)\frac{\partial h(x,z)}{\partial x}\right]+\frac{\partial}{\partial z}\left[K(x,z)\frac{\partial h(x,z)}{\partial z}\right]=0 \tag{31}$$

where $h(x, z)$ is the hydraulic head, that is equal to total head potential. The interstitial groundwater flow velocities, used in the governing transport Eq. (28), were determined using the hydraulic head distribution obtained by solving Eq. (31) in conjunction with the following equations:

$$U_x(x,z)=\frac{K(x,z)}{\theta}\frac{\partial h(x,z)}{\partial x} \tag{32}$$

$$U_x(x,z)=\frac{K(x,z)}{\theta}\frac{\partial h(x,z)}{\partial z} \tag{33}$$

where θ is the effective porosity of the porous medium. The porosity was assumed constant even though the hydraulic conductivity field is variable [51, 52]. The spatially variable hydrodynamic dispersion coefficients were calculated using the following relationships [50]:

$$D_x(x,z)=\frac{\alpha_T U_z^2(x,z)+\alpha_T U_x^2(x,z)}{|\mathbf{U}|}+\mathcal{D}_e \tag{34a}$$

$$D_z(x,z)=\frac{\alpha_T U_x^2(x,z)+\alpha_T U_z^2(x,z)}{|\mathbf{U}|}+\mathcal{D}_e \tag{34b}$$

where

$$|\mathbf{U}|=[U_x^2(x,z)+U_z^2(x,z)]^{1/2} \tag{35}$$

is the magnitude of the interstitial velocity vector. Local time dependent mass transfer coefficient values along the length of the NAPL pool were determined by modifying Eq. (5) so that $\bar{k}(t)$ can efficiently be evaluated numerically as follows:

$$\bar{k}(t)=\frac{1}{\ell_x N}\sum_{i=1}^{N}\int_{\ell_{x_o}}^{\ell_{x_o}+\ell_x} k_i(t,x)\,dx \tag{36}$$

where N is the number of different log-normally distributed hydraulic conductivity field realizations examined.

The contaminant transport model, Eq. (28), was solved using the backwards in time alternating direction implicit (ADI) finite difference scheme subject to a zero dispersive flux boundary condition applied to all outer boundaries of the numerical domain with the exception of the NAPL-water interface where concentrations were kept constant at the 1,1,2-TCA solubility limit C_s. The groundwater model, Eq. (31), was solved using an implicit finite difference scheme subject to constant head boundaries on the left and right of the numerical domain, and no-flux boundary conditions for the top and bottom boundaries, corresponding to the confining layer and impermeable bedrock, respectively, as schematically illustrated in Fig. 5. For each realization of the hydraulic conductivity field, local mass transfer coefficients were determined at each node at the NAPL-water interface from [24]:

$$\mathcal{D}_e \frac{\partial C(t, x, 0)}{\partial z} = -k(t, x)\, C_s \qquad \ell_{x_0} < x < \ell_{x_0} + \ell_x \tag{37}$$

by using the second-order accurate one-sided approximation of the concentration gradient above the NAPL pool [53]. The local mass transfer coefficients were averaged over the length of the NAPL pool and the average mass transfer coefficient was determined from Eq. (36) by averaging the results from the various realizations of the hydraulic conductivity field.

4.2
Numerical Model Simulations

The numerical model was employed to obtain aqueous phase concentrations originating from the dissolution of a 1,1,2-TCA pool in a heterogeneous, isotropic formation for two different variances of the log-transformed hydraulic conductivity distribution. The generated contours are presented in Fig. 6. Clearly, the results presented in Fig. 6 indicate that with increasing σ_Y^2 the heterogeneity of the porous medium is increased, the spreading of the dissolved plume is increased in the vertical direction, and consequently, the concentration gradients at the pool-water interface are reduced leading to $\bar{k}$ reduction.

Numerically evaluated average mass transfer coefficients, $\bar{k}$, based on the average of 200 different realizations of the log-normal hydraulic conductivity, as a function of ζ_z/ζ_x for several variances of $Y=\ln K$ ($\sigma_Y^2=0.1, 0.2, 0.3, 0.4$, and 0.5) for a hydraulic gradient of $\partial h/\partial x=0.01$ and mean log-transformed hydraulic conductivity of $\bar{Y}=0.8$ are shown in Fig. 7a. The results indicate that for increasing ζ_z/ζ_x there is a significant increase in $\bar{k}$. Low values of the anisotropy

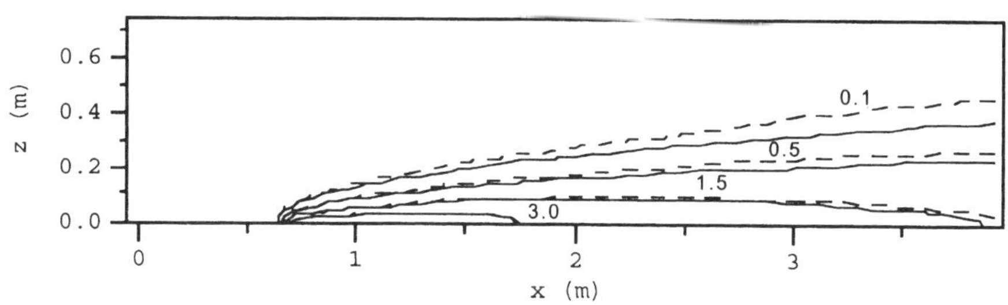

Fig. 6 Comparison between aqueous phase 1,1,2-TCA concentration contours originating from a 1,1,2-TCA pool for $\sigma_Y^2=0.5$ (*dashed line*) and 0.1 (*solid line*) (here $C_s=4.5$ g/l, $\mathcal{D}_e=2.33\times10^{-6}$ m^2/h, $\ell_{x_0}=0.64$ m, $\ell_x=0.72$ m, $R=1.63$, $t=5{,}000$ h, $\bar{Y}=0.8$, $\partial h/\partial x=0.01$, $\alpha_L=3.3\times10^{-2}$ m, $\alpha_T=3.3\times10^{-3}$ m, $\zeta_x=0.5$ m, $\zeta_z=0.5$ m, $\theta=0.3$) (adopted from Vogler and Chrysikopoulos [44])

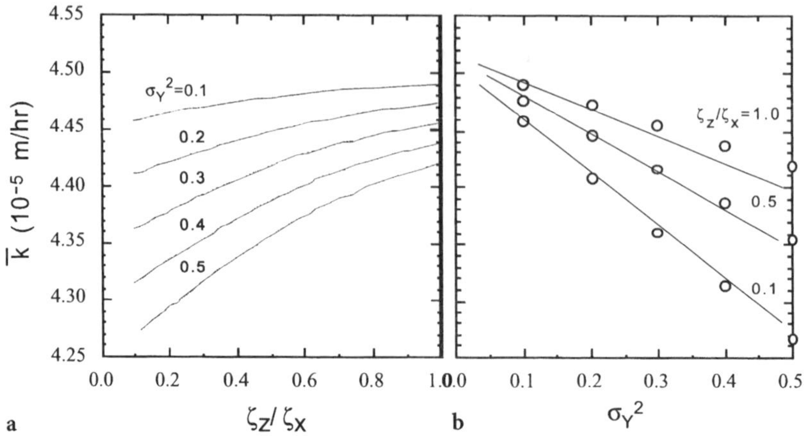

Fig. 7a, b Average mass transfer coefficient as a function of: **a** aquifer anisotropy ratio for several variances of the log-transformed hydraulic conductivity distribution; **b** variance of the log-transformed hydraulic conductivity distribution where *open circles* represent numerically generated data and *solid lines* represent linear fits. All model parameter values are identical with those used in Fig. 6

ratio ζ_z/ζ_x may represent the presence of thin lenticular layers of varying hydraulic conductivity which are observed in some sedimentary rocks [54]. Consequently, for low ζ_z/ζ_x ratios, transverse dispersion is significant causing more vertical spreading of the dissolved 1,1,2-TCA concentration plume which in turn leads to a thicker concentration boundary layer, smaller concentration gradients at the pool water interface, and smaller $\bar{k}$ values. Increasing ζ_z/ζ_x, the hydraulic conductivity becomes progressively more statistically similar in the vertical direction. For high ζ_z/ζ_x ratios the aqueous phase 1,1,2-TCA concentration boundary layer is thinner because vertical dispersion is reduced across statistically similar lenticular layers yielding steeper concentration gradients at the pool-water interface and greater $\bar{k}$ values.

Numerically evaluated $\bar{k}$ values, based on the average of 200 different realizations of the lognormal hydraulic conductivity, as a function of σ_Y^2 for several anisotropy ratios (ζ_z/ζ_x=0.1, 0.5, and 1.0) for a hydraulic gradient of $\partial h/\partial x$=0.01 and mean log-transformed hydraulic conductivity of $\bar{Y}$=0.8 are shown in Fig. 7b together with the best fitted lines. The results indicate that $\bar{k}$ is inversely proportional to σ_Y^2. Furthermore, in agreement with the results presented in Fig. 7a, it is also evident from Fig. 7b that $\bar{k}$ values increase with increasing ζ_z/ζ_x. However, increasing σ_Y^2 results in increased aquifer heterogeneity with a broader range of K values.

5 NAPL Pool Dissolution in the Presence of Dissolved Humic Substances

Humic substances are the most abundant, polydisperse organic materials in nature. They are formed by oxidation and condensation reactions between polyphenols, polysaccharides, and polyamino acids of plant and microbial origin [55]. Humic substances found in groundwater originate from infiltrating surface water, and from paleo-soils that were buried during geological sedimentation. The ability of humic substances to interact with hydrophobic organic compounds and to enhance hydrophobic organic solubility by a partitioning process is widely recognized [56]. However, the exact solubility enhancement mechanism is not well understood. The capability of dissolved humic substances to increase NAPL pool dissolution rates and enhance contaminant transport in a two-dimensional, homogeneous porous formation has been explored by Tatalovich et al. [57].

A two-dimensional numerical model was developed to describe the transport of dissolved organics originating from nonaqueous phase liquid pool dissolution in saturated porous media in the presence of dissolved humic substances. The model assumes that the dissolved contaminant (NAPL dissolved in the aqueous phase) may sorb onto the solid matrix as well as onto humic substances suspended in the aqueous phase, contaminant-humic particles and humic substances may sorb onto the solid matrix, the dissolved contaminant may undergo first-order decay, and humic substances are introduced into the aquifer from a line source.

5.1 Two-Dimensional Solute-Humic Cotransport Numerical Model

The appropriate mathematical model for solute transport originating from the dissolution of a perchloroethylene (PCE) pool located onto a bedrock within a two-dimensional, homogeneous, water saturated aquifer in the presence of dissolved humic substances, as illustrated in Fig. 8, consists of three coupled transport equations. One equation describing the transport of the solute originating from the dissolving PCE pool in the presence of dissolved humic substances, another equation describing the transport of dissolved humic substances, and the third equation describing the transport of solute-humic particles.

5.1.1 Solute Transport and Effective Mass Transfer

The two-dimensional transient transport of the solute originating from the dissolution of a PCE pool in saturated, homogeneous porous media under uniform interstitial groundwater flow and in the presence of dissolved humic substances is governed by [57]

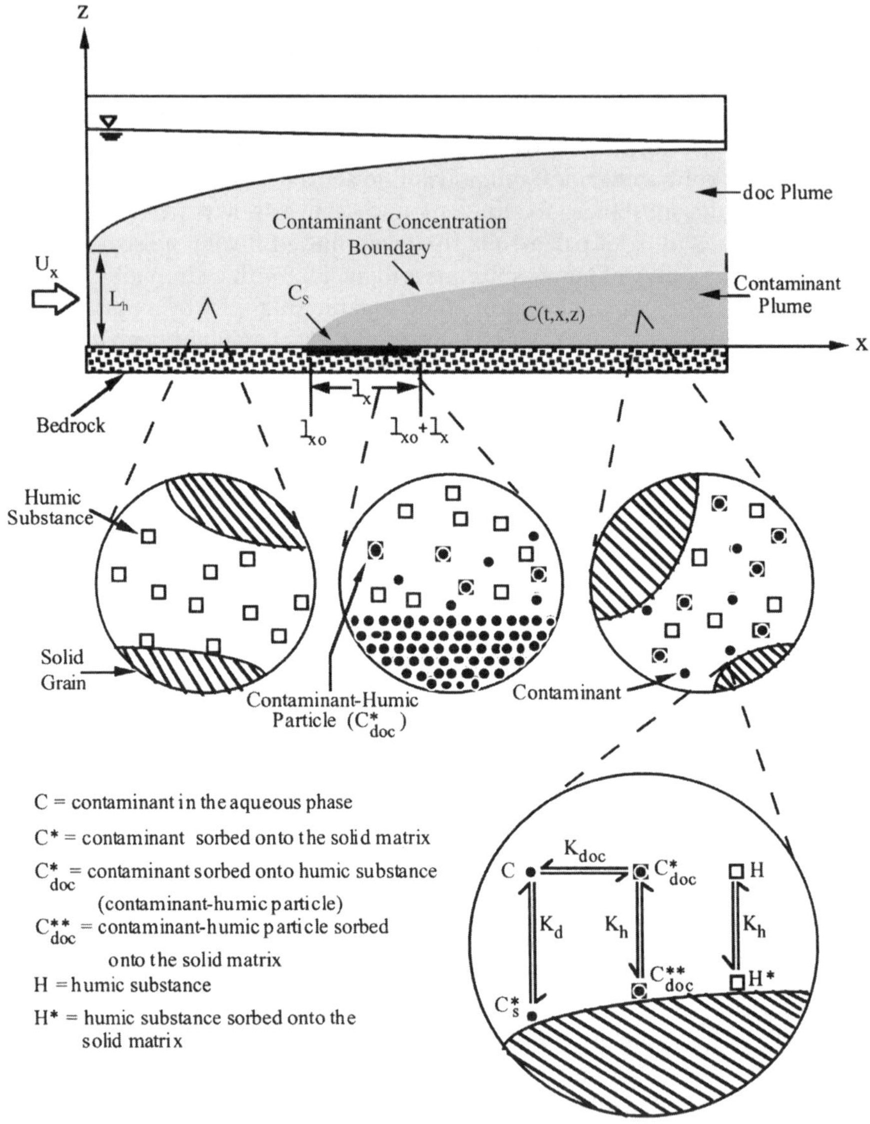

Fig. 8 Schematic diagram of the conceptual model showing the unidirectional velocity U_x, the DNAPL pool with aqueous saturation concentration C_s, the humic substance line source of height L_h, and the modes of interaction between dissolved contaminant, dissolved humic substances, and the solid matrix (adopted from Tatalovich et al. [57])

$$\frac{\partial}{\partial t}\left[R_s(t,x,z)\,C(t,x,z)\right] = D_{x_s}\frac{\partial^2 C(t,x,z)}{\partial x^2} + D_{z_s}\frac{\partial^2 C(t,x,z)}{\partial z^2} - U_x\frac{\partial C(t,x,z)}{\partial x} - \lambda_s R_s(t,x,z)\,C(t,x,z) \tag{38}$$

where the subscript s indicates the solute; and $R_s(t, x, z)$ is the time and space dependent, dimensionless retardation factor of the solute.

Assuming that there is no initial dissolved PCE concentration in the homogeneous, two-dimensional, water saturated formation of infinite dimension along the x direction and semi-infinite dimension along the z direction with an impervious boundary at $z=0$, as illustrated in Fig. 8, the appropriate initial and boundary conditions are

$$C(0,x,z) = 0 \tag{39}$$

$$\frac{\partial C(t,\pm\infty,z)}{\partial x} = \frac{\partial C(t,x,\infty)}{\partial z} = 0 \tag{40}$$

$$\frac{\partial C(t,x,0)}{\partial z} = \begin{cases} \dfrac{\partial C_a(t,x,0)}{\partial z} & \ell_{x_o} \le x \le \ell_{x_o} + \ell_x, \\ 0 & \end{cases} \tag{41}$$

otherwise, where $C_a(t, x, z)$ is the apparent (enhanced) aqueous phase solute concentration due to the presence of dissolved organic carbon (doc).

Numerous experimental studies suggest that a linear relationship exists between NAPL solubility and dissolved humic substance concentration [58–63]. Consequently, for a two-dimensional system, the apparent aqueous phase solute concentration at the NAPL-water interface can be expressed as [56]

$$C_a(t,x,z) = C_s\left[1 + K_{doc}\,H(t,x,0)\right] \tag{42}$$

where $H(t, x, z)$ is the dissolved organic carbon concentration attributable to humic substances; and K_{doc} is the solute partition coefficient between humic substances and the aqueous phase (based on the doc concentration). It should be noted that for $z\neq0$ the aqueous solubility, C_s, present in Eq. (42) should be replaced by the aqueous phase solute concentration, $C(t, x, z)$.

Assuming that the process of solute sorption onto the solid matrix can be described by the following linear equilibrium relationship:

$$C^*(t,x,0) = K_d\,C(t,x,z) \tag{43}$$

where $C^*(t, x, z)$ is the solute concentration sorbed onto the solid matrix, and that, in view of Eq. (42), the process of solute sorption onto dissolved humic substances is described by the following expression:

$$C^*_{\mathrm{doc}}(t,x,z) = K_{doc}\,C(t,x,z)\,H(t,x,z) \tag{44}$$

where $C^*_{\mathrm{doc}}(t, x, z)$ is the solute concentration sorbed onto dissolved organic carbon; consequently, the time and space dependent, dimensionless retardation factor of the solute, $R_s(t, x, z)$, is given by

$$R_s(t, x, z) = 1 + \frac{\rho_b K_d}{\theta} + K_{doc} H(t, x, z) \tag{45}$$

Given that the presence of humics enhance the solute solubility, the local mass transfer coefficient, k, defined in Eq. (3) should be modified accordingly. In view of Eq. (42), the appropriate expression for the effective local mass transfer coefficient, $\bar{k}_e(t, x)$, is given by [57]

$$\begin{aligned} k_e(t, x) &= -\frac{\mathcal{D}_e}{C_s}\frac{\partial}{\partial z}[C(t, x, 0) + C^*_{\mathrm{doc}}(t, x, 0)] \\ &= -\frac{\mathcal{D}_e}{C_s}\frac{\partial C_a(t, x, 0)}{\partial z} \end{aligned} \tag{46}$$

In the absence of dissolved humic substances, the preceding equation reduces to the local mass transfer coefficient defined in Eq. (3). Furthermore, the corresponding average effective mass transfer coefficient, $\bar{k}_e(t)$, applicable to the entire pool, can be expressed as

$$\bar{k}_e(t) = \frac{1}{\ell_x}\int_0^{\ell_x} k_e(t, x)\,\mathrm{d}x \tag{47}$$

5.1.2
Transport of Dissolved Humic Substances

The two-dimensional transient transport of dissolved humic substances in water saturated, homogeneous porous media under uniform interstitial groundwater flow is governed by the following partial differential equation:

$$R_h \frac{\partial H(t, x, z)}{\partial t} = D_{x_h}\frac{\partial^2 H(t, x, z)}{\partial x^2} + D_{z_h}\frac{\partial^2 H(t, x, z)}{\partial z^2} - U_x \frac{\partial H(t, x, z)}{\partial x} \tag{48}$$

where the subscript h indicates the dissolved humic substance; and R_h is the dimensionless retardation factor for humic substances. The governing transport Eq. (48) does not account for decay of humic substances because they are usually stable and persistent in the environment. Note that the dispersion coefficients for the solute-humic substances are differentiated from the dispersion coefficients for the solute because of their differences in molecular size and structure. Sorption of dissolved humic substances onto solids is often described mathematically as a linear equilibrium, Langmuir, or Freundlich process [64–67]. Here it is assumed that the sorption of dissolved humic substances onto the solid matrix is described by the following linear relationship:

$$H^*(t,x,z) = K_h\, H(t,x,z) \tag{49}$$

where $H^*(t,x,z)$ is the dissolved organic carbon concentration sorbed onto the solid matrix; and K_h is the humic substance linear equilibrium partition coefficient between the solid matrix and the aqueous phase. Consequently, the retardation factor, R_h, is given by

$$R_h = 1 + \frac{\rho_b\, K_h}{\theta} \tag{50}$$

Assuming that initially there are no dissolved humic substances in the homogeneous, two-dimensional, water saturated formation considered here, and the humic substances are introduced in the porous medium from a continuous line source of height L_h, located at x=0, y=0, as indicated in Fig. 8, the appropriate initial and boundary conditions are

$$H_0(0,x,z) = 0 \tag{51}$$

$$H(t,0,z \le L_h) = H_0 \tag{52}$$

$$\frac{\partial H(t,\pm\infty,z)}{\partial x} = \frac{\partial H(t,x,\infty)}{\partial z} = \frac{\partial H(t,x,0)}{\partial z} = 0 \tag{53}$$

where H_0 is the source concentration of dissolved humic substances expressed as doc.

5.1.3 Transport of Solute-Humic Substances

The two-dimensional, transient transport of solute-humic substances in water saturated, homogeneous porous media under uniform interstitial groundwater is governed by the following partial differential equation:

$$R_{sh}\frac{\partial C^*_{\mathrm{doc}}(t,x,z)}{\partial t} = D_{x_{sh}}\frac{\partial^2 C^*_{\mathrm{doc}}(t,x,z)}{\partial x^2} + D_{z_{sh}}\frac{\partial^2 C^*_{\mathrm{doc}}(t,x,z)}{\partial z^2} - U_x\frac{\partial^2 C^*_{\mathrm{doc}}(t,x,z)}{\partial x} - \lambda_{sh} R_{sh}\, C^*_{\mathrm{doc}}(t,x,z) \tag{54}$$

where subscript sh indicates the solute-humic substance; and R_{sh} is the dimensionless retardation factor for the dissolved solute-humic substances. Assuming that the sorption of dissolved solute-humic substances onto the solid matrix is described by the following linear relationship:

$$C^{**}_{\mathrm{doc}}(t,x,z) = K_h\, C^*_{\mathrm{doc}}(t,x,z) \tag{55}$$

where $C^{**}_{\mathrm{doc}}(t,x,z)$ is the solute-humic concentration sorbed onto the solid matrix, the solute-humic substances retardation factor, R_{sh}, is defined as

$$R_{sh} = R_h = 1 + \frac{\rho_b\, K_h}{\theta} \tag{56}$$

Note that the partition coefficient for solute-humic substances between the solid matrix and the aqueous phase, K_h, is identical to the partition coefficient for humic substances between the solid matrix and the aqueous phase. Furthermore, it was assumed that solute-humic substances share the same transport properties as humic substances, because humic substances are much larger than solutes and that sorption of nonpolar compounds onto humic substances most likely does not significantly affect the sorption characteristics of humic macromolecules. Also, it was assumed that the decay of solute-humic substances is dictated by the decay of the solute. Therefore, $D_{x_h}=D_{x_{sh}}$, $D_{z_h}=D_{z_{sh}}$, and $\lambda_{sh}=\lambda_s$.

The appropriate initial and boundary conditions for the transport of solute-humic substances in the two-dimensional, homogeneous, water saturated formation considered here are the following:

$$\partial C^*_{\mathrm{doc}}(0, x, z) = 0 \tag{57}$$

$$\frac{\partial C^*_{\mathrm{doc}}(t, \pm\infty, z)}{\partial x} = \frac{\partial C^*_{\mathrm{doc}}(t, x, \infty)}{\partial z} = \frac{\partial C^*_{\mathrm{doc}}(t, x, 0)}{\partial z} = 0 \tag{58}$$

5.2 Numerical Model Simulations

The numerical solution for the solute-humic cotransport model was obtained by an unconditionally stable, fully implicit finite difference discretization method. The three governing transport Eqs. (38), (48), and (54) in conjunction with the initial and boundary conditions given by Eqs. (39)–(41), (51)–(53), (58) and (59) were solved simultaneously [57]. All flux boundary conditions were estimated using a second-order accurate one sided approximation [53].

Model simulations were performed in order to examine the effect of humic substances on the effective average mass transfer coefficient. The behavior of $\bar{k}_e$, defined in Eq. (47), as a function of time in the presence and absence of humic substances is illustrated in Fig.9a. Note that the parameter L_h is set equal to zero when dissolved humic substances are not present. Clearly, the results indicate that the presence of humic substances increases the mass transfer from the PCE pool into the aqueous phase. An increase in the mass transfer coefficient leads to faster dissolution of the PCE pool.

To illustrate further the effects of humic substances on NAPL pool solubility enhancement, dissolved PCE concentrations as a function of time were determined for an arbitrary point within the aquifer with coordinates x=39.2 cm and z=1.0 cm. The results are presented in Fig. 9b. The total PCE concentration represents the sum of the PCE sorbed onto doc, C^*_{doc}, and the aqueous PCE concentration, C. The doc ($\times 10^{-1}$) concentration is also presented in Fig. 9b. As the dissolved humic substances pass over the PCE-water interface, PCE dissolution is enhanced due to sorption of the dissolved contaminant onto suspended humic substances. However, the aqueous phase contaminant concentration at the pool-water interface is assumed to remain equal to C_s. When contami-

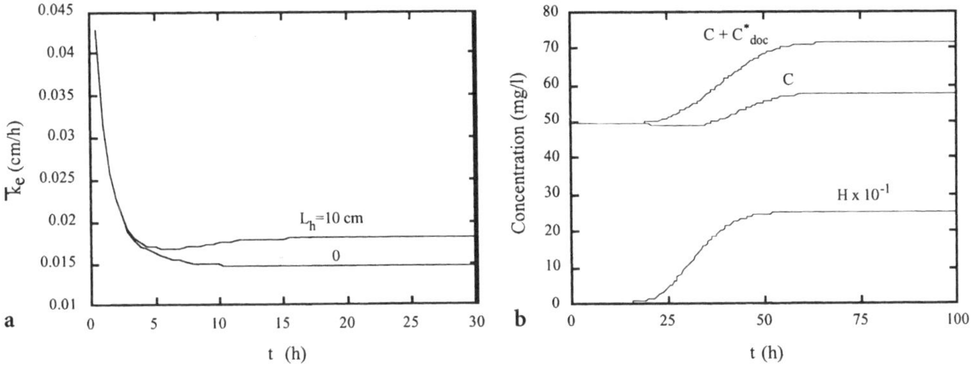

Fig. 9a, b Variation of the effective average mass transfer coefficient with increasing time evaluated for a PCE pool in the presence (L_h=10 cm) and absence (L_h=0 cm) of humic substances. **b** Breakthrough curves of the aqueous phase PCE concentration, C, total PCE concentration, $C+C^*_{doc}$, and doc concentration, H, at a point within the aquifer with coordinates x=39.2 cm, z=1 cm (here C_s=150 mg/l, $\mathcal{D}_{e_s}$=0.0219 cm^2/h, $\mathcal{D}_{e_h}$=0.009 cm^2/h, $\mathcal{D}_{x_s}$=1.222 cm^2/h, $\mathcal{D}_{x_h}$=1.209 cm^2/h, $\mathcal{D}_{z_s}$=0.622 cm^2/h, $\mathcal{D}_{z_h}$=0.609 cm^2/h, H_0=250 mg doc/l, K_d=0.310 l/kg dry solids, $K_{doc}=9.56\times10^{-4}$ l/mg doc, K_h=0.117 l/kg dry solids, ℓ_x=8.0 cm, ℓ_{x_o}=7.2 cm, L_h=10 cm, U_x=2.0 cm/h, α_x=0.6 cm, α_z=0.3 cm, ρ_b=1.47 g/cm^3, and θ=0.40)

nant-humic substances travel through sections of the aquifer with low aqueous phase contaminant concentrations, the contaminant desorbs from suspended humic substances into the aqueous phase, thus increasing the aqueous phase contaminant concentration. The concentration of PCE sorbed onto dissolved humic substances is the difference between the total PCE concentration and the aqueous phase PCE concentration.

6
Mass Transfer Correlations

Numerous empirical correlations for the prediction of residual NAPL dissolution have been presented in the literature and have been compiled by Khachikian and Harmon [68]. On the other hand, just a few correlations for the rate of interface mass transfer from single-component NAPL pools in saturated, homogeneous porous media have been established, and they are based on numerically determined mass transfer coefficients [69, 70]. These correlations relate a dimensionless mass transfer coefficient, i.e., Sherwood number, to appropriate Peclet numbers, as dictated by dimensional analysis with application of the Buckingham Pi theorem [71, 72], and they have been developed under the assumption that the thickness of the concentration boundary layer originating from a dissolving NAPL pool is mainly controlled by the contact time of groundwater with the NAPL-water interface that is directly affected by the interstitial groundwater velocity, hydrodynamic dispersion, and pool size. For uniform groundwater flow, the velocity boundary layer can be ignored. Therefore, fluid properties such as dynamic viscosities and densities for both groundwater and NAPL do not contribute to the formation of the concentration boundary layer.

The three-dimensional mathematical model presented by Eqs. (7), (10)–(12), (14) and

$$C(t,x,y,0) = C_s \qquad x,y \in \mathcal{R}_{(r)} \text{ or } \mathcal{R}_{(e)}, \tag{59a}$$

$$\frac{\partial C(t,x,y,0)}{\partial z} = 0 \qquad x,y \notin \mathcal{R}_{(r)} \text{ or } \mathcal{R}_{(e)} \tag{59b}$$

where $R_{(r)}$ and $R_{(e)}$ are the domains for rectangular and elliptic pool-water interfacial areas, defined as

$$\mathcal{R}_{(r)} : \ell_{x_o} \leq x \leq \ell_{x_o} + \ell_x \quad \text{and} \quad \ell_{y_o} \leq y \leq \ell_{y_o} + \ell_y, \tag{60}$$

$$\mathcal{R}_{(e)} : \frac{(x-\ell_{x_o})^2}{a^2} + \frac{(y-\ell_{y_o})^2}{b^2} \leq 1, \tag{61}$$

respectively, was solved numerically by an alternating direction implicit (ADI) finite-difference scheme. It should be noted that a circular pool with radius r is just a special case of an elliptic pool; therefore, the appropriate source boundary condition for a circular pool is obtained by setting $a=b=r$.

6.1 Local Mass Transfer Correlations

Chrysikopoulos and Kim [70] developed time invariant, local mass transfer correlations for NAPL pool dissolution in saturated media based on numerically determined local mass transfer coefficients evaluated for interstitial fluid velocities of 0.1, 0.5, 0.7, 0.85, and 1.0 m/day. It should be noted that solubility concentrations may occur at a NAPL-water interface when interstitial fluid velocities are less than 1.0 m/day [8]. Over 200 different rectangular pools with dimensions $\ell_x \times \ell_y$ in the range from 0.2 m×0.2 m to 10.0 m×10.0 m and approximately the same number of elliptic/circular pools with semiaxes $a \times b$ in the range 0.1 m×0.1 m to 5.0 m×5.0 m were examined.

6.1.1 Rectangular Pools

The fundamental parameters that affect mass transfer from a single component rectangular NAPL pool are the interstitial fluid velocity, effective molecular diffusion, dispersion coefficients, local coordinates of a specific location at the NAPL-water interface, and the square root of the rectangular pool area, $\ell_{c(r)}$, representing a characteristic length. The time invariant local mass transfer coefficient for a single component rectangular NAPL pool can be represented by the following functional relationship:

$$\hat{k}(x',y') = f[U_x, \mathcal{D}_e, D_x, D_y, x', y', \ell_{c(r)}], \tag{62}$$

where

$$x' = x - \ell_{x_o}, \tag{63}$$

$$y' = y - \left(\ell_{y_o} + \frac{\ell_y}{2}\right) \tag{64}$$

$$\ell_{c(r)} = (\ell_x \ell_y)^{1/2}, \tag{65}$$

f is an arbitrary function, and x' and y' represent the shifted coordinate system. Using dimensional analysis, Chrysikopoulos and Kim [70] have shown that for rectangular NAPL pools the local Sherwood number is related to appropriate local Peclet numbers as follows:

$$Sh_{(r)}(x',y') = \beta_1 \, Pe_{x(r)}^{\beta_2} \, Pe_{y(r)}^{\beta_3} \tag{66}$$

where

$$Sh_{(r)} = \frac{\hat{k}(x',y')\, x', y'}{\mathcal{D}_e \, \ell_{c(r)}} \tag{67}$$

$$Pe_{x(r)} = \frac{U_x x'}{D_x} \tag{68}$$

$$Pe_{y(r)} = \frac{U_x y'}{D_y} \tag{69}$$

$$\beta_1 = 0.01 \; \ell_x^{-0.53} (\ell_y/2)^{1.16} \; U_x^{-0.11}, \tag{70}$$

$$\beta_2 = 0.69 \; \ell_x^{0.13} \; U_x^{0.04}, \tag{71}$$

$$\beta_3 = 1.35 \, (\ell_y/2)^{-0.55} \; U_x^{0.01}, \tag{72}$$

where the subscript (r) designates a rectangular pool. The Sherwood number, $Sh_{(r)}$, is defined as the rate of interface mass transfer resistance to molecular diffusion resistance that physically represents the dimensionless concentration gradient at the NAPL-water interface. The local Peclet numbers, $Pe_{x(r)}$ and $Pe_{y(r)}$, represent the advective-dispersive mass transfer in the x and y directions, respectively. The correlation coefficients β_1, β_2, and β_3 were determined by fitting Eq. (66) to Sherwood numbers evaluated by numerically estimated $\hat{k}(x',y')$ values for various hydrodynamic conditions. It should be noted that unlike overall Sherwood number correlations with constant coefficients which are often published in the literature, Eq. (66) is a proposed empirical relationship for the local Sherwood number. Therefore, the local empirical coefficients are shown to depend on pool geometry and hydrodynamic conditions. The Eqs. (70)–(72) are valid for groundwater velocities in the range from 0.1 to 1.0 m/d and rectangular pools with dimensions $\ell_x \times \ell_y$ in the range from 0.2 m×0.2 m to 10.0 m×10.0 m.

6.1.2 Elliptic/Circular Pools

For a single component elliptic/circular NAPL pool the fundamental parameters that affect the local mass transfer coefficient are the same as those listed for the case of a rectangular pool. Therefore, the time invariant local mass transfer coefficient for a single component elliptic NAPL pool can be represented by the following functional relationship:

$$\hat{k}(x',y') = f[U_x, \mathcal{D}_e, D_x, D_y, x', y', \ell_{c(e)}], \tag{73}$$

where

$$x' = x - \ell_{x_o}, \tag{74}$$

$$y' = y - \ell_{y_o}, \tag{75}$$

$$\ell_{c(e)} = (\pi ab)^{1/2}, \tag{76}$$

x' and y' represent the coordinates of the Cartesian coordinate system with the origin shifted to the center of the elliptic pool, and $\ell_{c(e)}$ is the corresponding characteristic length that represents the square root of the elliptic pool area. Using dimensional analysis, Chrysikopoulos and Kim [70] have shown that for elliptic NAPL pools the local Sherwood number is related to appropriate local Peclet numbers as follows:

$$Sh_{(e)}(x',y') = \gamma_1 \, Pe_{x(e)}^{\gamma_2} \, Pe_{y(e)}^{\gamma_3}, \tag{77}$$

where

$$Sh_{(e)} = \frac{\hat{k}(x',y')\,|x'|\,y'}{\mathcal{D}_e \, \ell_{c(e)}} \tag{78}$$

$$Pe_{x(e)} = \frac{U_x |x'|}{D_x}, \tag{79}$$

$$Pe_{y(e)} = \frac{U_x \, y'}{D_y}, \tag{80}$$

$$\gamma_1 = 0.10\,(2a)^{-3.26}\, b^{-1.51}\, U_x^{-1.21}, \tag{81}$$

$$\gamma_2 = 5.31\,(2a)^{0.27}\, b^{-0.20}\, U_x^{0.21}, \tag{82}$$

$$\gamma_3 = 7.65\,(2a)^{0.10}\, b^{-0.19}\, U_x^{0.24}, \tag{83}$$

and the subscript (e) designates an elliptic/circular pool, and $|x'|$ is the absolute value of the distance from the center of the pool along the flow direction. The correlation coefficients γ_1, γ_2, and γ_3 were determined by fitting Eq. (77) to Sherwood numbers evaluated by numerically estimated $\hat{k}(x',y')$ values for various hydrodynamic conditions. The correlation presented can be applied to circular pools by setting $a=b=r$. The Eqs. (81)–(83) are valid for groundwater velocities in the range from 0.1 to 1.0 m/d and elliptic/circular NAPL pools with semiaxes $a \times b$ in the range from 0.1 m×0.1 m to 5.0 m×5.0 m.

6.2 Average Mass Transfer Correlations

Kim and Chrysikopoulos [69] developed time invariant, average mass transfer correlations for NAPL pool dissolution in saturated media, based on numerically determined average mass transfer coefficients evaluated for interstitial fluid velocities of 0.3, 0.5, 0.7, and 1.0 m/day. It should be noted that the dimensionless retardation factor R was arbitrarily set to 1.0, because at steady state conditions concentration profiles for a nondecaying solute are no longer dependent on R [24]. The dispersion coefficients D_x, D_y, and D_z were chosen in the range from 2.5–8.3×10^{-2} m^2/h, 2.5–8.3×10^{-3} m^2/h and 2.5–8.3×10^{-3} m^2/h, respectively. A total of 121 different rectangular pools with dimensions $\ell_x \times \ell_y$ in the range from 5.0 m×5.0 m to 10.0 m×10.0 m, and 121 different elliptic/circular pools with semiaxes $a \times b$ in the range 2.5×2.5 m to 5.0×5.0 m were examined.

6.2.1 Rectangular Pools

The time invariant, average mass transfer coefficient of a single component rectangular NAPL pool can be expressed by the following functional relationship:

$$k^{*} = f[U_x, \mathcal{D}_e, D_x, D_y, \ell_x, \ell_y, \ell_{c(\mathrm{r})}]. \tag{84}$$

Kim and Chrysikopoulos [69] have shown that for rectangular NAPL pools the average Sherwood number is related to appropriate average Peclet numbers as follows:

$$Sh^{*}_{(\mathrm{r})} = \beta_1 (Pe^{*}_{x(\mathrm{r})})^{\beta_2} (Pe^{*}_{y(\mathrm{r})})^{\beta_3}, \tag{85}$$

where

$$Sh^{*}_{(\mathrm{r})} = \frac{k^{*} \ell_{c(\mathrm{r})}}{\mathcal{D}_e}, \tag{86}$$

$$Pe^{*}_{x(\mathrm{r})} = \frac{U_x \ell_x}{D_x}, \tag{87}$$

$$Pe^{*}_{y(\mathrm{r})} = \frac{U_x \ell_y}{D_y}, \tag{88}$$

β_1, β_2, and β_3 are empirical coefficients that were determined by fitting the nonlinear power law correlation Eq. (85) to 484 average Sherwood numbers computed for 121 different pool dimensions and four different sets of hydrodynamic conditions. The resulting time invariant, average mass transfer correlation for rectangular pools is given by

$$Sh^{*}_{(r)} = 1.58\,(Pe^{*}_{x(r)})^{0.34}\,(Pe^{*}_{y(r)})^{0.43} \tag{89}$$

This correlation is valid for groundwater velocities in the range from 0.1 to 1.0 m/d, and rectangular NAPL pools with dimensions $\ell_x \times \ell_y$ in the range from 5.0 m×5.0 m to 10.0 m×10.0 m.

6.2.2 Elliptic/Circular Pools

The time invariant, average mass transfer coefficient of a single component elliptic NAPL pool can be expressed by the following functional relationship:

$$k^{*} = f[U_x, \mathcal{D}_e, D_x, D_y, \ell_x, \ell_y, \ell_{c(e)}]. \tag{90}$$

Kim and Chrysikopoulos [69] have shown that for elliptic NAPL pools the average Sherwood number is related to appropriate average Peclet numbers as follows:

$$Sh^{*}_{(e)} = \gamma_1\,(Pe^{*}_{x(e)})^{\gamma_2}\,(Pe^{*}_{y(e)})^{\gamma_3}, \tag{91}$$

where

$$Sh^{*}_{(e)} = \frac{k^{*}\,\ell_{c(e)}}{\mathcal{D}_e}, \tag{92}$$

$$Pe^{*}_{x(e)} = \frac{U_x\,a}{D_x}, \tag{93}$$

$$Pe^{*}_{y(e)} = \frac{U_x\,b}{D_y}, \tag{94}$$

γ_1, γ_2, and γ_3 are empirical coefficients that were determined by fitting the nonlinear power law correlation Eq. (91) to 484 average Sherwood numbers computed for 121 different pool dimensions and four different sets of hydrodynamic conditions. The resulting time invariant, average mass transfer correlation for elliptic pools is given by

$$Sh^{*}_{(e)} = 1.74(Pe^{*}_{x(e)})^{0.33}\,(Pe^{*}_{y(e)})^{0.40} \tag{95}$$

This correlation can be applied to circular pools by setting $a=b=r$. It should be noted that the Eq. (95) is valid for groundwater velocities in the range from 0.1 to 1.0 m/d, and elliptic/circular pools with semiaxes $a \times b$ in the range 2.5×2.5 m to 5.0×5.0 m.

7
Experimental Studies of NAPL Pool Dissolution in Porous Media

Chrysikopoulos et al. [73] designed and constructed a unique three-dimensional, bench-scale model aquifer to carry out DNAPL pool dissolution experiments. The model aquifer was constructed by glass with a specially designed aluminum bottom plate. A 0.5 cm deep, circular disk with 7.6 cm diameter was removed from the aluminum sheet in order to form the experimental pool. Glass beads were placed in the pool in order to support a #60 stainless steel mesh that prevented settling of sand into the pool. A 16 gauge stainless tube inserted into the pool through a horizontal hole drilled along the aluminum sheet was used to deliver the trichloroethylene (TCE) into the experimental pool. The aquifer material was well-characterized, uniform sand kiln-dried Monterey sand (RMC Lonestar, Monterey, California). The aqueous phase pumped through the experimental aquifer was degassed, deionized water with sodium azide. A schematic diagram of the three-dimensional, bench-scale model aquifer and a blowup of the sampling plate is shown in Fig. 10. It should be noted that a total

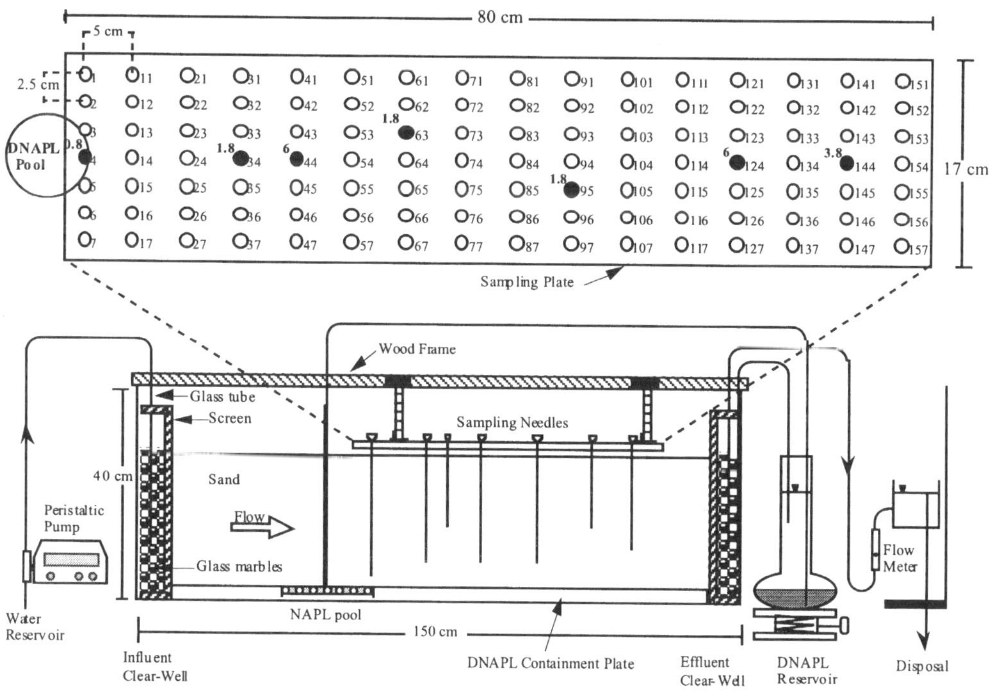

Fig. 10 Schematic diagram of the experimental aquifer and the sampling plate. *Filled circles* on the sampling plate represent needle placement locations. For each *filled circle*, the *lower right number* indicates the sampling port number and the *upper left bold numbers* indicate the needle placement depth measured in cm from the bottom of the aquifer (adopted from Lee and Chrysikopoulos [74])

of seven needles were vertically inserted into the aquifer and their locations are indicated on the sampling plate as filled circles. The designation of each sampling port is specified in Fig. 10 by a lower right number; whereas, the needle placement depth, measured in centimeters from the bottom of the aquifer, is indicated by a bold number at the upper left side of each sampling port.

The characteristics of the experimental aquifer were independently determined from appropriate flowthrough column experiments or obtained directly from the literature. The dry bulk density of the sand ρ_b=1.61 kg/l, and the aquifer porosity θ=0.415 were evaluated by gravimetric procedures. The dimensionless retardation factor, R=1.31, of the aqueous-phase TCE was determined from a column flowthrough experiment. The tortuosity coefficient for the aquifer sand was considered to be τ^*=1.43 [75]. The molecular diffusion coefficient for the aqueous-phase TCE is D=0.0303 cm^2/h [76]. The pool radius is r=3.8 cm. Bromide ion in the form of the moderately soluble potassium bromide salt was the tracer of choice [77] for the tracer experiment conducted in order to determine the longitudinal and transverse aquifer dispersivities α_L=0.259 cm and α_T= 0.019 cm, respectively. The experimental pool contained approximately 12 ml of certified ACS grade (Fisher Scientific) TCE with solubility of C_s=1100 mg/l [78].

7.1 TCE Dissolution Data

The experimental aqueous-phase TCE concentrations at specific locations within the aquifer downstream from the TCE pool for seven different interstitial fluid velocities (0.25, 0.51, 0.75, 1.21, 1.50, 1.96, and 3.35 cm/h) and steady-state conditions were collected by Chrysikopoulos et al. [73] and Lee and Chrysikopoulos [74]. For each TCE dissolution experiment, samples were simultaneously collected from sampling ports 4 (located at x=0.0, y=0.0, z=0.8 cm), 34 (15.0, 0.0, 1.8), 63 (30.0, 2.5, 1.8), 95 (45.0, –2.5, 1.8), and 144 (70.0, 0.0, 3.8). It should be noted that the origin of the Cartesian coordinate (0,0,0) is directly below port 4, and the center of the circular TCE pool is located at (–3.8,0,0). The interstitial velocity in the aquifer was altered by changing the water volumetric flow rate of the peristaltic pump. Aqueous phase TCE concentrations were collected only when steady state concentrations were observed at sampling port 144, that is the sampling port farthest away from the TCE pool (see Fig. 10). The duplicate aqueous phase TCE concentration experimental data were averaged and presented in Fig. 11. The experimental data indicate that aqueous phase TCE concentrations decrease with increasing U_x. This trend of the TCE concentration data was expected because the concentration boundary layer decreases with increasing interstitial velocity [30]. For each U_x considered, the highest dissolved concentration was observed at sampling port 4, that is closest to the TCE pool, and the lowest dissolved concentration was observed at node 144, that is the farthest away from the pool.

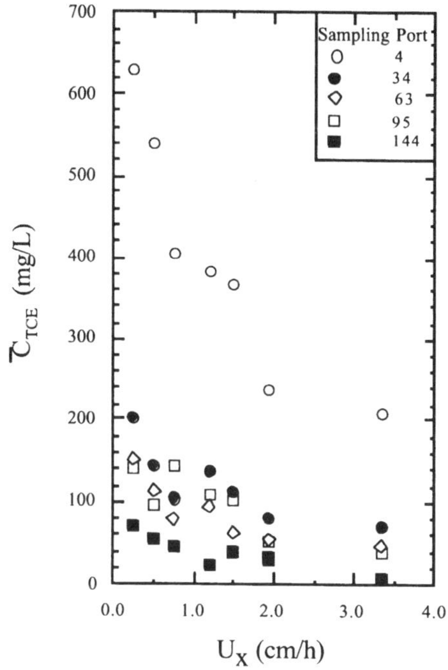

Fig. 11 Observed steady-state aqueous-phase TCE concentrations as a function of interstitial velocity at five different sampling ports (adopted from Lee and Chrysikopoulos [74])

7.2 Experimental Mass Transfer Correlation

For each set of aqueous-phase TCE concentrations collected, the corresponding k^* was estimated by fitting the analytical solution for a circular NAPL pool given by Eq. (21) to the experimental data with the non-linear least squares regression program PEST [79]. The estimated time invariant mass transfer coefficients together with the corresponding 95% confidence intervals, as determined by PEST, are presented in Fig. 12a. Clearly, it is shown that k^* increases steadily with increasing U_x. It should be noted that as U_x continues to increase, k^* approaches an asymptotic value.

The nonlinear least squares regression program PEST [79] was used to fit the proposed correlation relating the time invariant Sherwood number to overall Peclet numbers for circular pools given by Eq. (91) to the seven experimentally determined Sh^* values presented in Fig. 12b, in order to estimate the empirical coefficients γ_1, γ_2, and γ_3. The experimental, overall Sherwood number correlation applicable to circular TCE pool dissolution in water saturated, homogeneous porous media can be expressed by the following relationship:

$$Sh^*_{(e)} = 1.30\,(Pe^*_x)^{0.12}\,(Pe^*_y)^{0.44} \quad (96)$$

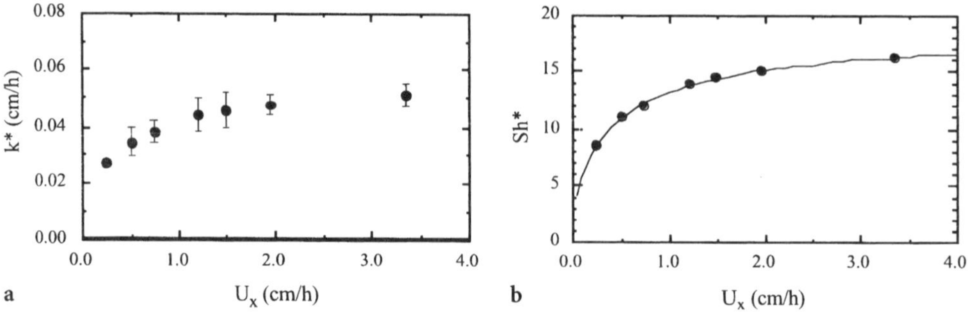

Fig. 12a, b Fitted time invariant overall mass transfer coefficients as a function of U_x evaluated from the averaged two sets of concentration measurements with error bars representing the 95% confidence intervals. **b** Comparison between experimentally determined Sh^* values (*solid circles*) and the experimental overall mass transfer correlation (*solid curve*) as a function of U_x

Figure 12b presents the experimentally determined correlation, Eq. (96), (solid curve) together with the experimentally determined Sh^* values (solid circles) as a function of interstitial velocity.

8 Summary

The fundamentals of mass flux at a NAPL-water interface within water saturated porous formations were discussed. Analytical solutions applicable for contaminant transport originating from NAPL pool dissolution in three-dimensional, homogeneous, water saturated porous media were presented for rectangular, elliptic as well as circular NAPL pools. The analytical solutions are useful for design and interpretation of experiments in laboratory-packed beds and possibly some homogeneous aquifers, and for the verification of complex numerical models. It was demonstrated through synthetic examples that the more elongated the pool along the direction of the interstitial flow, the higher the dissolved peak concentration. The effect of aquifer anisotropy and heterogeneity on the average mass transfer coefficient associated with DNAPL pool dissolution in water saturated porous media was examined with a numerical model. It was shown that increasing aquifer anisotropy results in an increase of the average mass transfer coefficient and that the average mass transfer coefficient is inversely proportional to the variance of the log-transformed hydraulic conductivity distribution. A two-dimensional finite difference numerical model was presented for the simulation of NAPL pool dissolution in a homogeneous, water saturated aquifer in the presence of dissolved humic substances. Model simulations indicated that dissolved humic substances may increase considerably NAPL pool dissolution. Local as well as average mass transfer correlations, based on numerical data, describing the rate of the interface mass transfer from single component rectangular or elliptic/circular NAPL pools in water saturated, homogeneous porous media were presented. The mass transfer correlations relate dimensionless mass transfer coefficients (Sherwood numbers) to appropriate Peclet numbers and can be used in existing analytical or numerical mathematical models simulating the transport of dissolved organics originating from the dissolution of NAPL pools in water saturated subsurface formations. Furthermore, an average mass transfer correlation based on experimentally determined mass transfer coefficients obtained from a unique, bench-scale, three-dimensional, TCE pool dissolution experiment was presented.

References

1. Schwille F (1984) Migration of organic fluids immiscible with water in the unsaturated zone. In: Yaron B, Dagan G, Goldshmid J (eds) Pollutants in porous media: the unsaturated zone between soil surface and groundwater. Springer, Berlin Heidelberg New York, pp 27–48
2. Mackay DM, Cherry JA (1989) Groundwater contamination: pump-and-treat remediation. Environ Sci Technol 23(6):630–636
3. Seagren EA, Rittmann BE, Valocchi AJ (1994) Quantitative evaluation of the enhancement of NAPL-pool dissolution by flushing and biodegradation. Environ Sci Technol 28:833–839
4. Schwille F (1981) Groundwater pollution in porous media by fluids immiscible with water. In: van Diujvenbooden W, Glasberger P, Lelyveld HH (eds) Quality of groundwater, studies in environmental science, vol 17. Elsevier Science, New York, NY, pp 451–463
5. Anderson MR, Johnson RL, Pankow JF (1992) Dissolution of dense chlorinated solvents into groundwater. 3. Modeling contaminant plumes from fingers and pools of solvent. Environ Sci Technol 26:901–908
6. Schwille F (1988) Dense chlorinated solvents in porous and fractured media. Translated by Pankow JF. Lewis, Chelsea, Michigan
7. Mackay DM, Roberts PV, Cherry JA (1985) Transport of organic contaminants in groundwater. Environ Sci Technol 19:364–392
8. Powers SE, Loureiro CO, Abriola LM, Weber WJ Jr (1991) Theoretical study of the significance of nonequilibrium dissolution of nonaqueous phase liquids in subsurface systems. Water Resour Res 27:463–477
9. Fried JJ, Muntzer P, Zilliox L (1979) Ground-water pollution by transfer of oil hydrocarbons. Ground Water 17:586–594
10. Abriola LM, Pinder GF (1985) A multiphase approach to the modeling of porous media contamination by organic compounds. 1. Equation development. Water Resour Res 21:11–18
11. Powers SE, Abriola LM, Weber WJ Jr (1992) An experimental investigation of nonaqueous phase liquid dissolution in saturated subsurface systems: seady state mass transfer rates. Water Resour Res 28:2691–2705
12. Thomson NR, Graham DN, Farquhar GJ (1992) One-dimensional immiscible displacement experiments. J Contam Hydrol 10:197–223
13. Lenhard RJ, Johnson TG, Parker JC (1993) Experimental observations of nonaqueous phase liquid subsurface movement. J Contam Hydrol 12:79–101
14. Pantazidou M, Sitar N (1993) Emplacement of nonaqueous liquids in the vadose zone. Water Resour Res 29:705–722

15. Imhoff PT, Jaffé PR, Pinder GF (1994) An experimental study of complete dissolution of a nonaqueous phase liquid in saturated porous media. Water Resour Res 30:307–320
16. Powers SE, Abriola LM, Weber WJ Jr (1994) An experimental investigation of nonaqueous phase liquid dissolution in saturated subsurface systems: Transient mass transfer rates. Water Resour Res 30:321–332
17. Bear J, Ryzhik V, Braester C, Entov V (1996) On the movement of an LNAPL lens on the water table. Transp Porous Media 25:283–311
18. Keller AA, Blunt MJ, Roberts PV (1997) Micromodel observation of the role of oil layers in three-phase flow. Transp. Porous Media 26:277–297
19. Kennedy CA, Lennox WC (1997) A pore-scale investigation of mass transport from dissolving DNAPL droplets. J Contam Hydrol 24:221–244
20. Jia C, Shing K, Yortsos YC (1999) Advective mass transfer from stationary sources in porous media. Water Resour Res 35:3239–3251
21. Brusseau ML, Nelson NT, Oostrom M, Zhang Z, Johnson GR, Wietsma TW (2000) Influence of heterogeneity and sampling method on aqueous concentrations associated with NAPL dissolution. Environ Sci Technol 34:3657–3664
22. Zhou D, Dillard LA, Blunt MJ (2000) A physically based model of dissolution of nonaqueous phase liquids in the saturated zone. Transp Porous Media 39:227-255
23. Johnson RL, Pankow JF (1992) Dissolution of dense chlorinated solvents into groundwater. 2. Source functions for pools of solvent. Environ Sci Tech 26:896–901
24. Chrysikopoulos CV, Voudrias EA, Fyrillas MM (1994) Modeling of contaminant transport resulting from dissolution of nonaqueous phase liquid pools in saturated porous media. Transp Porous Media 16:125–145
25. Voudrias EA, Yeh MF (1994) Dissolution of a toluene pool under constant and variable hydraulic gradients with implications for aquifer remediation. Groundwater 32:305–311
26. Lee KY, Chrysikopoulos CV (1995) Numerical modeling of three-dimensional contaminant migration from dissolution of multicomponent NAPL pools in saturated porous media. Environ Geol 26:157–165
27. Lee KY, Chrysikopoulos CV (1998) NAPL pool dissolution in stratified and anisotropic porous formations. J Environ Eng ASCE 124:851–862
28. Holman H-YN, Javandel I (1996) Evaluation of transient dissolution of slightly water-soluble compounds from a light nonaqueous phase liquid pool. Water Resour Res 32: 915–923
29. Mason AR, Kueper BH (1996) Numerical simulation of surfactant-enhanced solubilization of pooled DNAPL. Environ Sci Technol 30:3205–3215
30. Chrysikopoulos CV, Lee KY (1998) Contaminant transport resulting from multicomponent phase liquid pool dissolution in three-dimensional subsurface formations. J Contam Hydrol 31:1–21
31. Seagren EA, Rittmann BE, Valocchi AJ (1999) An experimental investigation of NAPL pool dissolution enhancement by flushing. J Contam Hydrol 37:111–137
32. Leij FJ, van Genuchten MT (2000) Analytical modeling of nonaqueous phase liquid dissolution with Green's functions. Transp Porous Media 38:141–166
33. Leij FJ, Priesack E, Schaap MG (2000) Solute transport modeled with Green's functions with application to persistent solute sources. J Contam Hydrol 41:155–173
34. Sciortino A, Harmon TC, Yeh W-G (2000) Inverse modeling for locating dense nonaqueous pools in groundwater under steady flow conditions. Water Resour Res 36: 1723–1735
35. Parker JC, Katyal AK, Kaluarachchi JJ, Lenhard RJ, Johnson TJ, Jayaraman K, Ünlü K, Zhu JL (1991) Modeling multiphase organic chemical transport in soils and groundwater. US Environmental Protection Agency Project CR-814320, Washington D.C., USA

36. Pinder GF, Abriola LM (1986) On the simulation of nonaqueous phase organic compounds in the subsurface. Water Resour Res 22:109s–119s
37. van der Waarden M, Bridie ALAM, Groenewoud WM (1971) Transport of mineral oil components to groundwater. 1. Model experiments on the transfer of hydrocarbons from a residual oil zone to trickling water. Water Res 5:213–226
38. Borden RC, Kao C-M (1992) Evaluation of groundwater extraction for remediation of petroleum-contaminated aquifers. Water Environ Res 64:28–36
39. Borden RC, Piwoni MD (1992) Hydrocarbon dissolution and transport-A comparison of equilibrium and kinetic models. J Contam Hydrol 10:309–323
40. Chrysikopoulos CV (1995) Three-dimensional analytical models of contaminant transport from nonaqueous phase liquid pool dissolution in saturated subsurface formations. Water Resour Res 31:1137–1145
41. Incropera FP, DeWitt DP (1990) Fundamentals of heat and mass transfer. Wiley, New York, NY, p 919
42. Bear J, Verruijt A (1987) Modeling groundwater flow and pollution. D. Reidel, Dordrecht, Holand, p 414
43. Hashimoto I, Deshpande KB, Thomas HC (1964) Peclet numbers and retardation factors for ion exchange columns. Ind Eng Chem Fund 3:213–218
44. Vogler ET, Chrysikopoulos CV (2001) Dissolution of nonaqueous phase liquid pools in anisotropic aquifers. Stoch Environ Res Risk Assess 15:33–46
45. Gutjhar AL (1989) Fast Fourier transform for random field generation. Los Alamos Report 4-R58-2690R, New Mexico Institute of Mining and Technology, Socorro, NM
46. Gelhar LW, Axness CL (1983) Three-dimensional stochastic analysis of macrodisisperion in aquifers. Water Resour Res 19:161–180
47. Sudicky EA (1986) A natural gradient experiment on solute transport in a sand aquifer: spatial variability of hydraulic conductivity and its role in the disperion process. Water Resour Res 22:2069–2082
48. Russo D, Zaidel J, Laufer A (1994) Stochatic analysis of solute transport in partially saturated heterogeneous soils. 1. Numerical experiments. Water Resour Res 30:769–779
49. Burr DT, Sudicky EA, Naff RL (1984) Nonreactive and reactive solute transport in three-dimensional heterogeneous porous media: mean displacement, plume spreading, and uncertainty. Water Resour Res 30:791–815
50. Bear J (1979) Hydraulics of goundwater. McGraw-Hill, New York, NY
51. Freeze RA (1975) A stochastic-conceptual analysis of one-dimensional groundwater flow in nonuniform homogeneous media. Water Resour Res 11:725–741
52. Dagan G (1989) Flow and transport in porous formations. Springer, Berlin Heidelberg New York
53. Strikwerda JC (1989) Finite difference schemes and partial differential equations. Wadsworth & Brooks/Cole, Pacific Grove, CA
54. Smith L (1981) Spatial variability of flow parameters in a stratified sand. Math Geol 13:1–21
55. Larson RA, Weber EJ (1994) Reaction mechanisms in environmental organic chemistry. CRC Press, Raton, FL
56. Chiou CT, Malcolm RL, Brinton TI, Kile DE (1986) Water solubility enhancement of some organic pollutants and pesticides by dissolved humic and fulvic acids. Environ Sci Technol 20:502–508
57. Tatalovich ME, Lee KY, Chrysikopoulos CV (2000) Modeling the transport of contaminants originating from the dissolution of DNAPL pools in aquifers in the presence of dissolved humic substances. Transp Porous Media 38:93–115
58. Carter CW, Suffet IH (1982) Binding of DDT to dissolved humic materials. Environ Sci Technol 16:735–740

59. Chiou CT, Kile DE, Brinton TI, Malcolm RL, Leenheer JA (1987) A comparison of water solubility enhancements of organic solutes by aquatic humic materials and commercial humic acids. Environ Sci Technol 21:1231–1234
60. Chin Y-P, Weber WJ Jr (1989) Estimating the effects of dispersed organic polymers on the sorption of contaminants by natural solids. 1. A predictive thermodynamic humic substance-organic solute interaction mode. Environ Sci Technol 23:978–984
61. Abdul AS, Gibson TL, Rai DN (1990) Use of humic acid solution to remove organic contaminants from hydrogeologic systems. Environ Sci Technol 24:106–114
62. Kopinke R-D, Porschmann J, Stottmeister U (1995) Sorption of organic pollutants on anthropogenic humic matter. Environ Sci Technol 29:941–950
63. Tanaka S, Oba K, Fukushima M, Nakayasu K, Hasebe K (1997) Water solubility enhancement of pyrene in the presence of humic substances. Anal Chim Acta 337:351–357
64. Murphy EM, Zachara JM, Smith SC, Phillips JL (1992) The sorption of humic acids to mineral surfaces and their role in contaminant binding. Sci Total Environ 117:413–423
65. Liljestrand HM, Lo I, Shimizu Y (1992) Sorption of humic materials onto inorganic surfaces for the mitigation of facilitated pollutant transport processes. Water Sci Technol 26:1221–1228
66. Jardine PM, Dunnivant FM, Selim HM, McCarthy JF (1992) Comparison of models for describing the transport of dissolved organic carbon in aquifer columns. Soil Sci Soc Am J 56:393–401
67. McCarthy JF, Williams TM, Liang L, Jardine PM, Jolley LW, Taylor DL, Palumbo AV, Cooper LW (1993) Mobility of natural organic matter in a sandy aquifer. Environ Sci Technol 27:667–676
68. Khachikian C, Harmon TC (2000) Nonaqueous phase liquid dissolution in porous media: Current state of knowledge and research needs. Transp Porous Media 38:3–28
69. Kim T-J, Chrysikopoulos CV (1999) Mass transfer correlations for nonaqueous phase liquid pool dissolution in saturated porous media. Water Resour Res 35:449–459
70. Chrysikopoulos CV, Kim T-J (2000) Local mass transfer correlations for nonaqueous phase liquid pool dissolution in saturated porous media. Transp Porous Media 38:167–187
71. Bird RB, Stewart WE, Lightfoot EN (1960) Transport Phenomena. Wiley, New York, NY
72. Weber WJ Jr, DiGiano FA (1996) Process dynamics in environmental system. Wiley, New York, NY
73. Chrysikopoulos CV, Lee KY, Harmon TC (2000) Dissolution of a well-defined trichloroethylene pool in saturated porous media: experimental design and aquifer characterization. Water Resour Res 36:1687–1696
74. Lee KY, Chrysikopoulos CV (2002) Dissolution of a well-defined trichloroethylene pool in saturated porous media: experimental results and model simulations. Water Res 36:3911–3918
75. de Marsily G (1986) Quantitative hydrogeology, groundwater hydrology for engineers. Academic Press, San Diego, CA, pp 440
76. Hayduk W, Laudie H (1974) Prediction of diffusion coefficients for nonelectrolytes in dilute aqueous solutions. AIChE J 20:611–615
77. Chrysikopoulos CV (1993) Artificial tracers for geothermal reservoir studies. Environ Geology 22:60–70
78. Mackay D, Shiu WY, Ma KC (1992) Illustrated handbook of physical-chemical properties and environmental fate for organic chemicals, volatile organic chemicals, vol 3. Lewis, Chelsea, MI, pp 916
79. Doherty J, Brebber L, Whyte P (1994) PEST: Model-independent parameter estimation. Watermark Computing, Brisbane, Australia

Handb Environ Chem Vol. 5, Part F, Vol. 3 (2005): 133–157
DOI 10.1007/b11435

A Case Study in the Application of Environmental Chemodynamic Principles for the Selection of a Remediation Scheme at a Louisiana Superfund Site

Raghava R. Kommalapati[1] (✉) · W. David Constant[2] · Kalliat T. Valsaraj[3]

[1] Department of Civil Engineering, PO Box 4249, Prairie View A&M University, Prairie View, TX 77446, USA
r_kommalapati@pvamu.edu

[2] Department of Civil and Environmental Engineering, Louisiana State University, Baton Rouge, LA 70803, USA

[3] Gordon A. and Mary Cain Department of Chemical Engineering, Louisiana State University, Baton Rouge, LA 70803, USA

1	**Introduction**	135
1.1	Site History	135
1.2	Literature	137
2	**Experimental Results and Discussion**	139
2.1	Soil-Water Partition Coefficients	139
2.2	Desorption from Laboratory-Spiked Soils and Adsorption/Desorption Hysteresis	140
2.3	Desorption from Field Contaminated Soil	144
2.4	Desorption Kinetics	146
2.5	Bioavailability Studies	149
2.5.1	Freshly Contaminated Soil	149
2.5.2	Desorption-Resistant Fraction	150
3	**Implications for Remediation at the PPI Sites**	151
References		155

Abstract This chapter studies the application of chemodynamic principles to select a remediation scheme at a Louisiana Superfund site (Petro Processors, Inc (PPI) sites). The current remediation scheme at the sites, monitored natural attenuation (MNA) is a direct result of this and other research conducted at Louisiana State University. In this chapter, the results from our studies on the adsorption and desorption, the desorption kinetics, the delineation of freely desorbing and desorption resistant fractions, and the bioavailability of the desorption resistant fraction are presented along with the implication of this research for the current remediation scheme.

We observed that only a small fraction of the adsorbed mass would desorb even after a number of successive desorption steps. The investigation on laboratory contaminated soil showed a biphasic behavior, namely an easily desorbed fraction and a desorption resistant fraction. Both field contaminated and aged soils also showed the same behavior. The first stage involved a "loosely bound" fraction and the second stage involved a "tightly bound" fraction. The desorption constants calculated or estimated for the two fractions were employed to obtain the overall expected mass recovery from the contaminated zone at the site. Extremely large time frames were predicted for overall mass removal of Hexachlorobutadiene (HCBD) from the contaminated zone.

An empirical non-linear model was used to describe the bi-phasic nature of desorption with one fraction (labile) being released in relatively short periods of time (typically 24–100 h) and a second fraction (non-labile or irreversible) being resistant to desorption and the parameters estimated. In addition, desorption kinetics of three-month and five-month old contaminated soils showed that progressively less amount of contaminant was available for labile desorption (lower F) compared to freshly contaminated soil.

We observed that for freshly contaminated soil, the compound readily desorbed into the aqueous phase and was available for microbial consumption whereas for soils containing mostly the non-labile material, the contaminant availability was limited by the mass transfer into the aqueous phase. The fraction of contaminant, which is irreversibly bound to soil is typically present in micropores or chemically bound to soil humic matter and thus is not accessible for microbial utilization. These observations are in agreement with those reported for other chemicals in the literature. It is believed that the longer the contaminant age within the soil the lower the fraction of the contaminant that will be bioavailable. The observations have significant implications to the current remedy and the possibility of natural attenuation at the site.

Keywords Natural attenuation · Hysteresis · Adsorption · Desorption · Biodegradation · Kinetics · Remediation · Chlorinated organics · Partition coefficients

List of Abbreviations

CD	Consent Decree
C_w	Aqueous concentration of the chemical
C_w^o	Initial aqueous concentration of the chemical
DCB	Dicholorobenzene
EAE	Environmentally acceptable end points
ε	Soil porosity
F	Fraction of chemical released quickly or the labile phase
f_{min}	Fraction of mineral matter in soil
f_{oc}	Fraction of organic carbon in soil
HOC	Hydrophobic Organic Compound
HCBD	Hexachlorobutadiene
k_1	First-order rate coefficient for the labile fraction
k_2	First-order rate coefficient for the non-labile fraction
K_d	Soil-water partition coefficient
K_{min}	Partition coefficient for the compound on mineral surface
K_{oc}	Organic carbon normalized partition coefficient
K_{ow}	Octanol-water partition coefficient
K_{sw}	Partition coefficient
$K_{sw}^{des,1}$	Partition coefficient for the loosely bound fraction
$K_{sw}^{des,2}$	Partition coefficient for the tightly bound fraction
K_{sw}^{rev}	Partition coefficient if partitioning was entirely reversible
L	Length of the zone
LADEQ	Louisiana Department of Environmental Quality
LSU	Louisiana State University
MNA	Monitored Natural Attenuation
n	Number of subdivisions in a zone
NPC	NPC Services Inc.

PPI	PetroProcessors, Inc.
ρ_b	Bulk density of soil
R_F	Retardation factor
S_a	Mineral surface area
S_t/S_o	Fraction of chemical released after time t
SRAP	Supplemental Remedial Action plan
τ_c	Chemical residence time
t	Time
V_x	Groundwater velocity
x	Distance
W_{Total}	Mass of total material adsorbed
W_{irr}	Mass of material adsorbed in the irreversible compartment
W_{rev}	Mass of material adsorbed in the reversible compartment

1 Introduction

1.1 Site History

The Petro Processors Inc. (PPI) sites comprise two former petrochemical disposal areas situated about 1.5 miles apart near Scotlandville, about 10 miles north of Baton Rouge, Louisiana, the Scenic Highway and Brooklawn sites, totaling 77 acres. Brooklawn is the larger of the two areas, currently estimated at 60 acres. These sites were operated in late 1960s and early 1970s to accept petrochemical wastes. During their operation, approximately 3.2×10^5 tons of refinery and petrochemical wastes were disposed in non-engineered pits on the two sites. Free phase organics are present in buried pits and high permeability soil lenses are found in the proximity of the disposal area. Contaminants at the sites are predominantly chlorinated organic solvents and aromatic hydrocarbons. These sites are on the United States Environmental Protection Agency (USEPA) National Priorities List. The companies involved with the waste disposal formed NPC Services, Inc. (NPC) to manage the remediation of the sites. A Supplemental Remedial Action Plan (SRAP) was approved in August 1989 as result of the Consent Decree (CD) entered into by NPC, the US EPA and Louisiana Department of Environmental Quality (LADEQ) in the US District Court, Middle District of Louisiana [1]. The amended consent decree of 1989: (i) provided the frame work to amend the initial remedy, which was to cover the sites with a clay cap and install a system of wells as part of a hydraulic containment/ recovery facility for free phase organic liquid and (ii) also directed College of Engineering at Louisiana State University (LSU) to conduct additional investigations on available treatment methods for remediation of the two PPI sites. The conventional hydraulic containment and recovery system was initiated with a plan for a total of 214 wells at the bigger Brooklawn site. However, this method requires long times to make significant reductions in the quantity of organic contaminants [2]. The removal of hydrophobic organic compounds (HOCs) from contaminated soils is usually hindered by very low solubility in water and high interfacial tension [3]. Thus, there is a need for new alternative technologies or other methods to enhance the recovery of contaminants. The aerial photographs of one of the two sites, Brook lawn, depicting how the site has evolved on the surface over the time is shown in Fig. 1.

Several research groups from LSU conducted feasibility studies on numerous innovative treatment technologies including; composting, electrokinetic remediation, engineered biodegradation (bioremediation), surfactant aided pump and treat technology, wherein surfactants in the form of either conventional aqueous solutions or colloidal gas aphron (CGA) suspensions were investigated to enhance recovery from the ongoing pump and treat wells [4]. After initial feasibility studies, surfactant technology and engineered bioremediation were considered for site demonstration. Our group was involved with the research

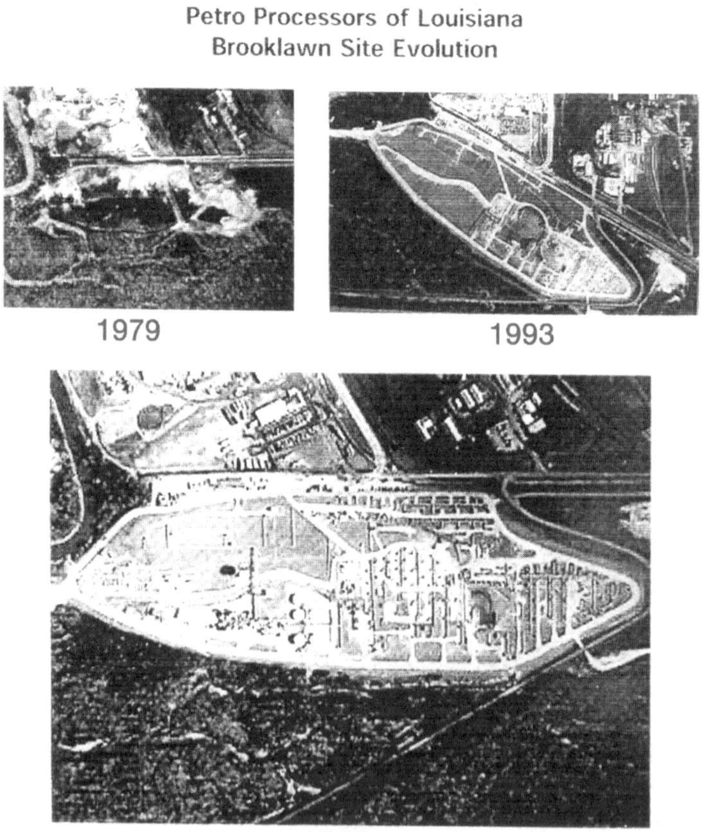

Fig. 1 Aerial view showing the evolution of Petro Processors Inc. (PPI) site (Brooklawn site) over the last 30 years

on both technologies and was instrumental in supporting the remedial actions at the sites. The use of surfactants to enhance the performance of the existing well facilities at Brooklawn was of significant interest to NPC, since the pump and treat wells were already in production, and any additional wells and equipment necessary could easily be incorporated into the system.

Surfactants could be introduced into the subsurface, either as conventional surfactant solutions or as colloidal gas aphron (CGA) suspensions. CGA suspensions have been observed to sweep contaminated soil more efficiently than water flood or conventional surfactant solutions during the flushing of laboratory soil columns [5]. The success of these technologies at the laboratory and pilot scale showed great promise for the surfactant technology. However, one potential problem that the researchers and the regulators were concerned about was the possibility of downward migration of mobilized contaminants and surfactants into deeper depths of the pristine subsurface soils. At about the same time these discussions were going on, results from our other research and new findings in the literature supported monitored natural attenuation for the two sites and the surfactant site demonstration research was terminated.

Our other research project, which was initiated to support the modeling activities by providing the estimates of soil-water partition coefficients for the harbinger compounds at the site for different soils to be used with the groundwater geochemical model, MODFLOW, yielded interesting findings. It was found that a significant fraction of the contaminant was undergoing irreversible adsorption and was not readily desorbed under normal conditions. However, data on the desorption kinetics of readily desorbing and desorption resistant fractions and the bioavailability of the desorption resistant fraction was not readily available in the literature. To fill this knowledge gap, our group undertook a comprehensive study on the sorption/desorption hysteresis, desorption kinetics and bioavailability of key contaminants found at the sites. The findings of these studies along with other supporting projects and recent literature on the subject eventually resulted in the selection of monitored natural attenuation (MNA) as the current remediation scheme now in place for the two sites.

This chapter details the findings of the sorption/desorption hysteresis and desorption kinetic studies and how the results of these environmental chemodynamic studies impacted remediation of the PPI sites. The literature on the topic is presented below.. In the next section, we present some of the results of our research on sorption/desorption hysteresis and desorption kinetics for freshly contaminated soils as well as aged soils. The experimental protocols are published elsewhere [6–10].

1.2 Literature

Hydrophobic organic compounds (HOCs), such as halogenated hydrocarbons are one of the most important classes of environmental pollutants at the PPI sites. MODFLOW and other advanced groundwater models are used to determine the trigger levels of key contaminants, which indicate sufficient containment/disappearance being maintained, thus preventing future degradation of the 400-foot aquifer as required by the Consent Decree [4]. Site geology and aquifer characteristics vary widely with the sands and clays present, and thus the partitioning of HOCs to the soils varies tremendously from zone to zone at the sites. It is, therefore, necessary to obtain best estimates of the soil-water partition coefficients, and have an understanding of the reversibility of the adsorption process.

Fate and transport and risk assessment models all contain terms for sorption and an understanding of the dynamics of sorption is crucial for correct prediction. Ignoring slow kinetics can lead to an underestimation of the true extent of sorption, false predictions about the mobility of contaminants and perhaps the wrong choice of cleanup or remediation technology options. The slow sorption of the contaminants onto soil leads to a chemical fraction that is resistant to desorption. Recent reports have suggested that desorption has a major slow release fraction after a comparatively rapid release of material occurs[7, 11]. The bi-phasic nature of desorption could significantly impact fate model predictions. It has been also reported in the literature that field contaminated soils exhibit slower grain scale desorption than found in laboratory spiked soils [12]. The resistant fraction for the aged soils, where contact times may have been months or years, can be enriched in the slow fraction owing to partial dissipation or degradation of more labile fractions before collection [13]. This is presumed to be due to sequestration of the contaminants within the soil and this could reduce the bioavailability to microorganisms in soil [14]. Studies suggested a two stage (bi-phasic) desorption of organic chemicals from soils and sediments wherein a rapid release of a labile fraction is followed by the slow release of a non-labile fraction [13, 15, 16]. Quantitative models have been only partly successful in explaining desorption hysteresis, irreversibility and slowly reversible, non-equilibrium behavior. However, two-site empirical models such as the one suggested by Opdyke and Loehr [17] seem to be better suited to describe the data on desorption. Chemical desorption from soil determines contaminant transport and mobility which are key factors in their fate assessment in the sub-surface. If sorption is kinetically limited, the partition coefficient, K_d, obtained may be vastly over-predicted and can introduce substantial uncertainties in assessing the transport of contaminants in groundwater moving away from a site.

Recent research has shown that HOCs, particularly aromatic compounds, may be biodegraded by microorganisms to a residual concentration that no longer decreases or which decreases only very slowly over years [18]. It is widely believed that further reductions are limited by the availability of hydrocarbons to microorganisms and all the more so for aged contaminants [19, 20]. The bioavailability of a chemical is controlled by a number of physical-chemical processes such as sorption and desorption, diffusion, and dissolution [21]. In particular, for aged soils a fraction of the contaminant appears to be inaccessible for biodegradation. The pollutant, not nutrient availability, seems to be the cause for the persistence of these chemicals. This may result from: (i) chemical oxidation reactions incorporating the contaminants into soil organic matter, (ii) slow diffusion into very small pores and adsorption into organic matter, or (iii) the formation of semi-rigid films around non-aqueous phase liquids with a high resistance toward organic-water mass transfer [19].

Several researchers have confirmed that biodegradation can be limited by the slow desorption of organic compounds [22–25]. Though significant research has been conducted to study the sorption and desorption kinetics of organic compounds and their bioavailability, few studies have focused on the bioavailability of contaminants in soils containing only the desorption resistant fraction and how the degradation rates compare to those for freshly contaminated soils.

To address some of the specific issues noted above as they relate to the PPI sites, a systematic environmental chemodynamic study was undertaken to study the fundamental mechanisms involved and provide a basis for the remediation scheme (MNA) that is being adopted for the two sites. The components of the study included (i) determining the soil-water partition coefficients for four harbinger compounds, (ii) studying the adsorption/desorption hysteresis using freshly contaminated soils as well as field contaminated soils, (iii) determining the kinetic parameters from the desorption kinetic studies for freshly contaminated and aged soils, (iv) studying the bioavailability of reversibly bound and desorption resistant fractions of the contaminants, and (v) interpreting the results to identify the implications of these findings on MNA as a remedy at the PPI sites. Details of the experimental methods can be found elsewhere [6–10].

2 Experimental Results and Discussion

2.1 Soil-Water Partition Coefficients

Sorption of hydrophobic compounds occurs primarily via partitioning of the non-polar solute out of the polar aqueous phase and onto the hydrophobic surfaces of the soil. The hydrophobic surfaces are usually associated with soil organic matter, i.e., the more hydrophobic an organic molecule is, the greater its incompatibility with water and thus the greater the sorption of the solute on the hydrophobic surface of the soil. The octanol-water partition coefficient thus is an excellent indicator of the hydrophobicity of a molecule. It is well recognized that not only the organic carbon present in the soil but also the mineral surfaces contribute to overall sorption of a compound. A two-phase model for sorption onto mineral surfaces and organic matter partitioning is proposed to account for this behavior, the soil-water partition coefficient, K_d is approximated as [26, 27]

$$K_d = f_{oc} K_{oc} + f_{min} S_a K_{min} \quad (1)$$

where f_{min} is the fraction mineral matter in the soil, S_a (m^2/Kg) is the mineral surface area, and K_{min} (L/m^2) is the partition coefficient for the compound on mineral surfaces, f_{oc} is the fraction of organic carbon in the soil and K_{oc} is the organic-carbon normalized partition coefficient for the compound. In order to obtain the contributions of both organic carbon and mineral matter, Eq. (1) is used in the following form:

$$\frac{K_d}{f_{oc}} = K_{oc} + K_{\min} S_a \left(\frac{f_{\min}}{f_{oc}}\right) \quad (2)$$

Hence a plot of K_d/f_{oc} for a specific compound and $S_a f_{min}/f_{oc}$ for the different types of soils would provide estimates for K_{min} (l/m^2) (slope) and K_{oc} (l/kg) (intercept).

The soil-water partition coefficients were obtained using three different soils from different locations from the PPI sites at several depths with four harbinger compounds (1,4-dichlorobenzene, 1,2 dichloroethane, 1,1,2,2-tetrachloroethane, and 1,1,2-trichloroethane). The soils provided a broad horizon and chemicals are major ones of concern at the sites. Average values were calculated from triplicate samples and used in plots mentioned above to determine log K_{oc} and log K_{min}, which were correlated to log K_{ow} to arrive at the general predictive equation for PPI soils given below [7]:

$$K_d = 10^{0.81} K_{ow}^{0.56} f_{oc} + 10^{-6.51} K_{ow}^{0.71} S_a f_{\min} \quad (3)$$

The correlations for K_{oc} are very good and are in good agreement with the literature. However the correlations for log K_{min} had a large 95% confidence level band indicating high variability. The predicted values from the above equation are in good agreement with those measured for the four test-compounds using a premixed composite soil, indicating the effectiveness of the predictive equation. The primary use of the above equation is within the groundwater model for the sites. In order to use the model for the heterogeneous geological strata at the site, one sets up a series of horizontal and vertical grids. The partition coefficients at each node are then input in the model. The above equation is convenient since only two soil parameters (f_{oc} and S_a) besides the compound specific parameter, K_{ow} are needed. A computer code for the above equation is easily incorporated as a subroutine into the existing model that is being currently used to predict MNA of the contaminant plumes.

2.2 Desorption from Laboratory-Spiked Soils and Adsorption/Desorption Hysteresis

In order to understand further the behavior for the four harbinger compounds as to their adsorption and desorption pathways, and to test the validity of reversible sorption, a series of experiments were performed whereby the adsorption step was followed by successive desorption using distilled water. The same four compounds (1,4-dichlorobenzene, 1,2-dichloroethane, 1,1,2-trichlorethane, and 1,1,2,2-tetrachloroethane) and a Pleistocene clay were used for the studies. Figure 2a shows the results of the adsorption/desorption experiment using 1,4-DCB as the test compound on Pleistocene clay soil. Data for the three other chemicals are not presented here, but the results are very similar. The line in the figure describes the linear adsorption isotherm for 1,4-DCB obtained using only the points during the adsorption process. The desorption isotherms obtained for 1,4-DCB are also shown in Fig. 2a for three initial spike concentrations. The data are also plotted as the total mass remaining on the soil vs. a function of both time and the number of desorption steps in Fig. 2b. It is clear that a significant fraction of 1,4-DCB remains on the surface even after several successive desorption steps.

Depending on the initial spike concentration, the aqueous concentrations were reduced to about 5 µg/l in 7 steps (11 days) for lowest spike concentration of 1 mg/l to 40 steps (160 days) for highest concentration 60 mg/l. The mass removed during desorption was roughly 70%. Thus, roughly 30% of the mass was not desorbed even after 40 desorption steps. The percent of desorption resistant fractions for each of the other three test chemicals at different initial soil concentration ranges were determined to range from 30 to 70%. This is within the range of values reported by several other authors [11, 28].

If adsorption and desorption were perfectly reversible, one should have expected over 99% recovery of material from the adsorbed state in five steps in all the cases. It is clear from the data in Fig. 2 that the total mass of compound remaining on the solid is dependent on the initial mass available for desorption. The mass of material remaining on the surface can be characterized as 'slowly desorbing'. A number of investigators have reported recently that similar significant 'slowly desorbing' or irreversible fractions has been observed in many other systems, including both freshly contaminated sediments and soils such as the present case and field aged sediments (for example see Kan et al. [11] and Pignatello and Xing [13]). The desorption of contaminants is said to follow a biphasic pattern. The reversible fraction desorbs easily within a short period of time, followed by the slow removal of the slowly desorbing fraction, which typically takes weeks to months. It is believed that the second compartment containing the slowly desorbing fraction is of a finite size and that once this compartment is filled to capacity, the additional mass that adsorbs does so reversibly [11]. The present experiments appear to corroborate this observation. The total mass of material adsorbed, W_{Total}, can be apportioned between the two compartments to give

$$W_{Total} = W_{rev} + W_{irr} \tag{4}$$

In order to obtain the irreversible fraction from the data shown in Fig. 2a, one re-plots the data on a linear scale as shown in Fig. 3a. The first four successive desorption data are linearly extrapolated to the Y-axis to obtain W_{irr} and the value of W_{rev} at each aqueous concentration was obtained by the difference using Eq. (4). A plot of W_{rev} vs the corresponding aqueous phase concentration can then be made to obtain the partition coefficient for the reversible fraction, K_d (Fig. 3b). A very good correlation of these data was obtained (r^2=0.99). The

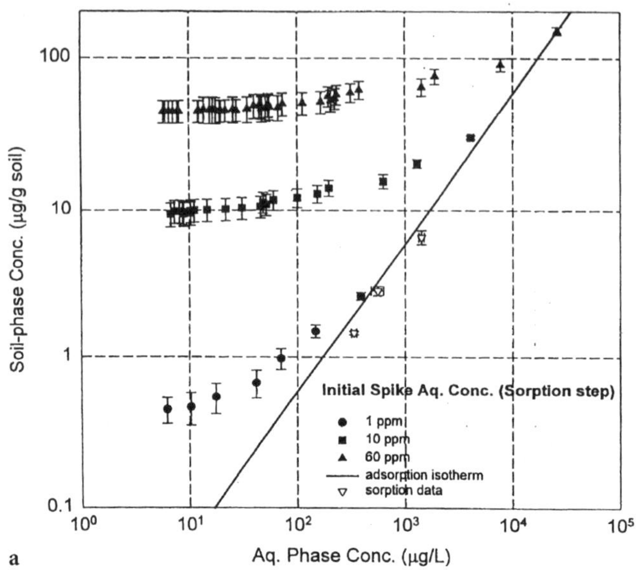

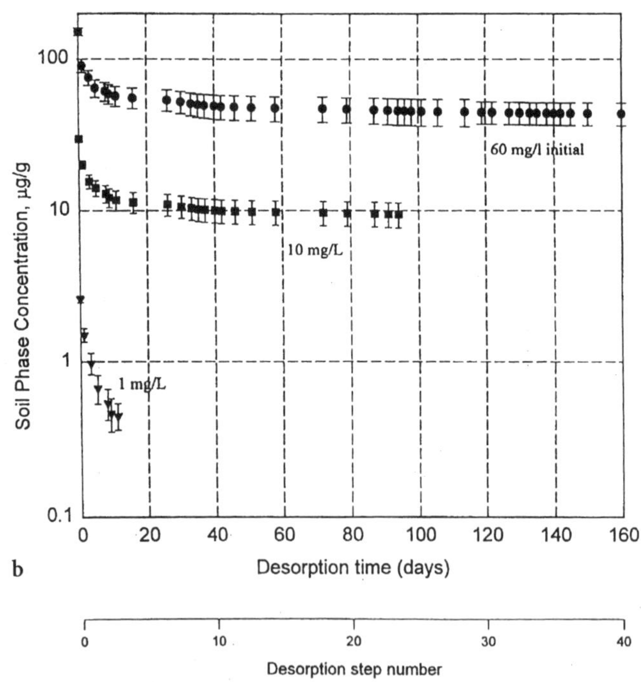

Fig. 2a, b a Adsorption and desorption of 1,4-DCB from Pleistocene clay (PC) soil. **b** Total mass of 1,4-DCB remaining on the soil as a function of both desorption time and number of steps during the desorption cycle. The *open square* symbol denotes the values calculated from the adsorption data

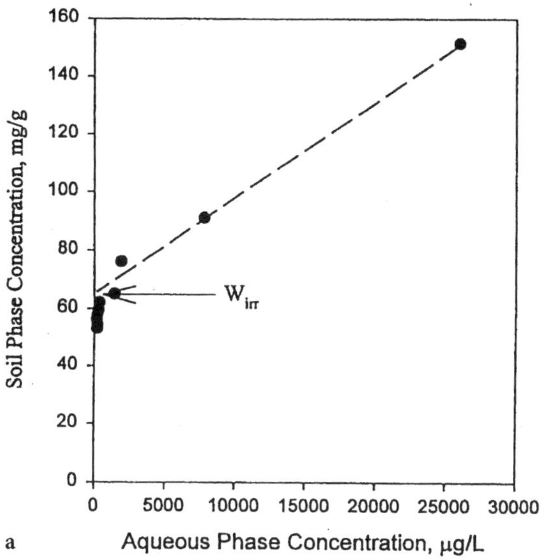

a

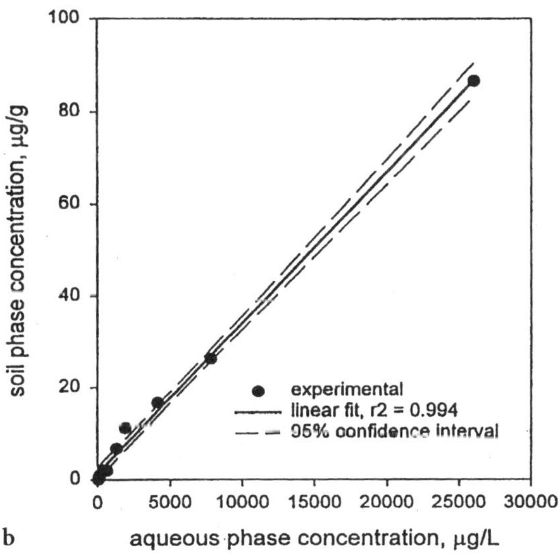

b

Fig. 3a, b **a** Plot of observed solid phase concentration vs aqueous phase concentration for 1,4-DCB; the *dashed line* represents the fit to the four initial points, the y-intercept of which is taken to be the irreversible fraction on the soil. **b** Plot of W_{rev} vs aqueous phase concentration

value of K_d obtained was 3.3 l/kg ($K_{oc}=10^{3.12}$) which is comparable to that obtained earlier during the adsorption step ($K_d=5.8$ l/kg and a corresponding $K_{oc}=10^{3.3}$). Two important observations can be made here: (i) a small fraction of the HOC resists desorption, and (ii) the fraction that adsorbs to the reversible compartment follows a linear isotherm with K_d obtained from the adsorption data. What stands out about these experiments is that adsorption occurred over a short time (24–48 h) on an uncontaminated soil, whereas the desorption was incomplete even after several weeks. This observation begs the question: what would happen to an already contaminated weathered soil at the PPI sites and what fraction of the material is likely to leach from the soil during the hydraulic containment and recovery remedy? The next few sections address this issue.

2.3 Desorption from Field Contaminated Soil

Long term persistence in soils of intrinsically biodegradable compounds, even when other environmental factors are not limiting microbial growth, supports the hypothesis advanced by Pignatello and Xing [13] and Hatzinger and Alexander [29] that the field contaminated (aged) soil, where contact times are months to years, can be enriched in the slowly desorbing fraction owing to partial dissipation or degradation of more labile fractions before collection. To test this hypothesis, we conducted an experiment using field contaminated soil from the PPI sites. The soil was chosen such that there were only trace levels of contaminants and no free phase liquids present. Though the soil contained several chlorinated organics the discussion here is limited to HCBD as it is one of the most prevalent compounds present at high concentrations throughout the site. It was found at a high initial concentration in soil, 1414±1 μg/g. The aqueous concentration after each desorption step is plotted in Fig. 4a. The initial concentration in water in equilibrium with the soil was 2120±114 μg/l, but the concentration did not decline as rapidly as was observed for the other three compounds (not shown here). The aqueous concentration declined to about 1200 μg/l after 20 desorption steps (140 days). The mass remaining on the soil after each desorption step was determined through a mass balance and the resulting desorption isotherm relating the aqueous concentration to the equilibrium soil concentration was plotted as shown in Fig. 2b. Each point on the plot represents the mean of quadruplicate samples.

If sorption was entirely reversible, an isotherm can be obtained by connecting the origin to the first desorption point. The slope of the line (667 l/kg) represents the partition coefficient, K_{sw}^{rev} if the partitioning was entirely reversible. It is quite clear that the adsorption and desorption pathways are quite different, and that considerable hysteresis in desorption exists. A linear fit of data in the desorption cycle yields a slope equivalent to a partition coefficient for the loosely bound fraction which we represent as $K_{sw}^{des,1}$ and is 166 l/kg in the present case. Extrapolating back on this slope suggests that approximately 1,204±13 μg/g of the HCBD is tightly bound to the soil and may desorb only very slowly (months

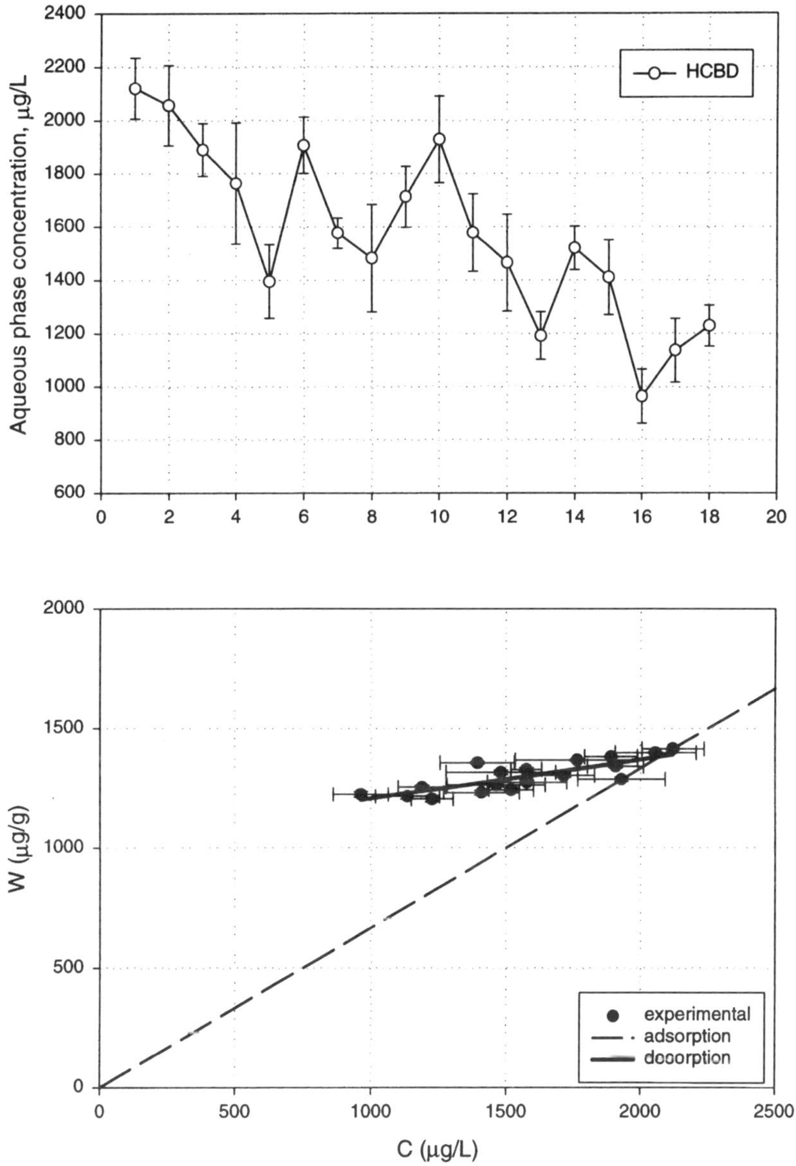

Fig. 4a, b a Aqueous phase concentration of HCBD during successive desorption steps. b Desorption isotherm for HCBD from the contaminated PPI site soil sample

to years). Thermodynamics dictate that the desorption curve cannot, in reality, extend to a finite value on the y-axis, and that at some concentration near the y-axis, a sharp decrease in soil concentration should be observed. An attempt was made to obtain the desorption constant for the tightly bound fraction by extending a line drawn from the origin to intersect the extrapolated line as shown in Fig. 4b. The slope obtained is $K_{sw}^{des,2}$=(1414–1204)/20=10.5 l/g=10,500 l/kg. Kan et al. [30] reported that the aqueous concentration of dichlorobenzene, *p,p'*-DDT and polychlorinated biphenyls in equilibrium with the maximum "irreversibly" sorbed compartment concentration on three different sediments was in the range of 1 to 20 µg/l. Hence, we selected the value of 20 µg/l as the aqueous concentration for calculation of the slope. This indicates a 160-fold decrease in HCBD concentration relative to its aqueous solubility. Thus, an estimated ratio of $K_{sw}^{des,2}/K_{sw}^{des,1}$ for HCBD on PPI site soil is 63. This is only meant to illustrate the relative magnitude of the two compartments into which the HCBD partitions within the soil. Gess and Pavlostathis [31] reported that the ratio for HCBD on an estuarine sediment from Bayou d'Inde in Louisiana was 33. A number of investigators have also shown that the ratio of the partition coefficients varied from 1 to 20 [13] for a variety of other compounds on contaminated soils.

2.4 Desorption Kinetics

Desorption kinetic studies were conducted with freshly contaminated soils and aged soils (i.e., soils that were allowed to have extended contact time of 3 months and 5 months during the sorption step). Three levels of contamination were used. To illustrate here, soil desorption kinetics data for 1,3-DCB with silty soil from the PPI site was plotted with the fraction of the contaminant released as a function of time in Fig. 5. Again, the results for other chemicals are not discussed here as the findings are very similar. As shown in Fig. 5, a substantial portion of the contaminant is released within the first 20–30 h, followed by a very slow release over a very long period. This slow release was observed over the entire duration of the experiment (100–450 h). Approximately 60% of 1,3-DCB was desorbed in the first 24 h for freshly contaminated soils.

An empirical model was used by Opdyke and Loehr [17] to describe the contaminant rate of release (ROR) similar to what we observed; a relatively rapid release of the chemical followed by a much slower release of the remaining chemical. The non-linear equation used to describe this biphasic behavior during desorption was given by

$$\frac{S_t}{S_0} = 1 - \mathrm{F}e^{-k_1 t} - (1 - F)e^{-k_2 t} \qquad (5)$$

where t is time, S_t/S_0 is the fraction of chemical released after time t. F is the fraction of chemical released quickly (labile phase) and (1–F) is the fraction of chemical released slowly (non-labile phase). k_1 and k_2 are the first order rate

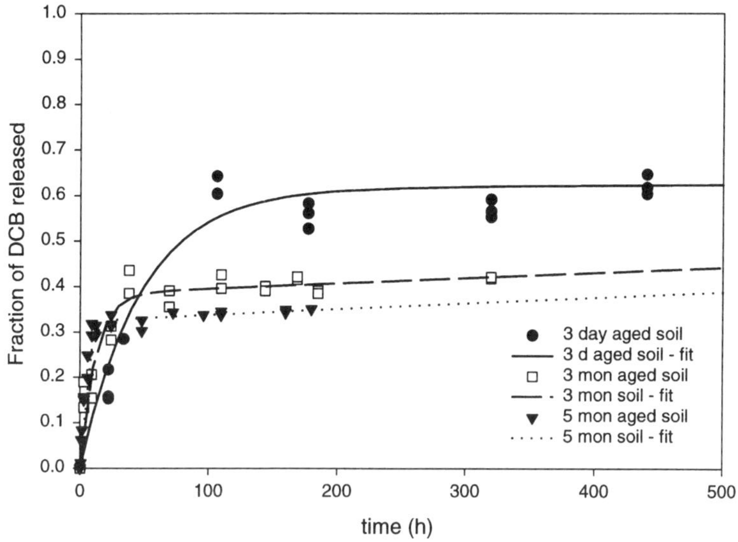

Fig. 5 The desorption kinetic data for DCB. It shows the effect of aging of contaminant within the soil on the desorption behavior of 1,3-DCB data for three different contaminant ages (3 days, 3 months, and 5 months) within the soil are shown The *lines* were obtained using the model parameters determined using the curve fit

constants describing the desorption of the labile fraction and slowly released (non-labile) fraction ($time^{-1}$). The model parameters, F, k_1, and k_2 were determined by fitting the experimental data to the model. The lines in the figures are obtained using Eq. (5) with the model parameters determined from the non-linear fit of the composite data. The first order rate coefficient for the labile (k_1) and non-labile (k_2) fractions are 0.03 and $6\times10^{-4}\ h^{-1}$ respectively. Thus, at least two orders of magnitude difference was found in the values of rate constants for the labile and non-labile fractions. The fraction of contaminant that is rapidly released (F=0.6 for DCB) provides an indication of the amount of organic contaminant that is available for immediate transport or uptake by microorganisms in the soil. The distribution of the contaminant among the labile and non-labile fractions has been found to be dependent on the type of chemical, soil organic carbon content and other factors including aging [7, 15, 32–34].

It was clear that an equilibrium desorption time of more than 100 h is required for 1,3-DCB. Accordingly, three triplicate samples each of the PPI soil at three different initial soil concentrations were kept well-mixed with 40 ml of distilled water for 180 h. The desorption isotherm thus obtained is plotted in Fig. 6. It is clear that this desorption isotherm indicates that a portion of the compound remains bound to the soil. The linear isotherm representing this would consist of: (i) a reversible portion obeying the conventional linear isotherm, $W_{rev}=K_{sw}C_w$, where W_{rev} is the soil concentration (μg/g), K_{sw} is the partition coefficient (l/g), and C_w is the aqueous phase concentration (μg/l), and (ii) a desorption

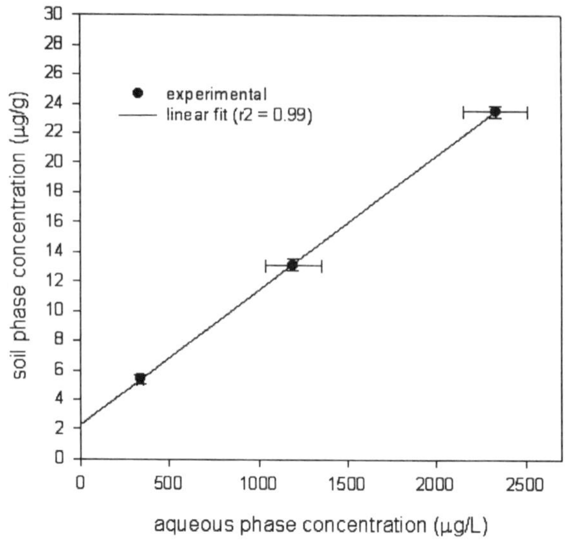

Fig. 6 Desorption isotherm 1,3-DCB using silty soil from the PPI sited

resistant component W_{irr} (µg/g). Thus $W=W_{irr}+K_{sw}C_w$. A plot of W vs C_w will give W_{irr} as the intercept and K_{sw} as the slope. For 1,3-DCB the value of W_{irr} obtained was 9.1 l/kg, which corresponded to a K_{oc} of 674 l/kg. The reported K_{oc} from adsorption measurements was 293 l/kg [35]. Many others [24, 36] also have reported that desorption partition coefficients are much larger than adsorption partition coefficients.

The desorption kinetics data presented in Fig. 5 also show the effect of the age of contamination for 1,3-DCB. The values of the model parameters, F, k_1, and k_2 are 0.38, 0.09, and 2×10^{-4} for three-month-old contamination and 0.32, 0.22, and 2×10^{-4} 2×10^{-4} for the five-month-old contamination. As expected, the labile fraction was reduced from about 60 to 32% for 1,3-DCB. The longer incubation period in the case of the aged soil allows organic molecules to diffuse and sorb onto the compartments or regions in the sorbent that exhibit slow adsorption/desorption kinetics [37–39]. However, the desorption rate constant, k_1, for the labile fraction was higher for the aged contaminated soil than the freshly contaminated soil. This suggests that the initial release from the aged soil is faster than the initial release from freshly contaminated soil. Therefore, during adsorption, all the reversible sites on the soil and the organic carbon are first occupied by the contaminant before it starts to accumulate in the irreversible sites. For freshly contaminated soil, within the short equilibrium time of three days, the contaminant should be mainly in the reversible sites and diffusion of contaminant into irreversible sites might still be occurring. Thus, both processes of desorption into water and migration to the non-labile sites are occurring simultaneously and the overall desorption rate is lowered. However, for aged soils, the diffusion of contaminant from the reversible sites to non-labile sites would be nearly complete in the three- and five-month periods. Thus, when the desorption process was initiated, the fraction of contaminant in the labile sites, though smaller, desorbs at a faster rate than from the freshly contaminated soil. This suggests that the 72-h equilibration time used in this experiment is not long enough for the contaminant to diffuse into the soil organic matter and micropores and thus larger fractions would be released when subjected to desorption. The aged soil study also further reinforces the need to assess F with k_1 and k_2 when comparing fast and slow release rates, as the rate coefficients also may be misleading if the aging process is not well known.

2.5 Bioavailability Studies

2.5.1 Freshly Contaminated Soil

Soil for these experiments was obtained by equilibrating an uncontaminated soil with an aqueous solution spiked with 1,3-DCB followed by separation of the aqueous phase. The soil thus obtained was analyzed for DCB concentration and used in the microcosms directly. The soil was amended with nutrients and inoculated with seed cultures. Again, the findings from preliminary studies were used to design the experiments to avoid losses from the system and also to determine the intervals for sampling. Figure 7 shows the percent of 1,3-DCB degraded as a function of incubation time for freshly contaminated soil and soil containing a desorption resistant fraction. Again, data points are obtained by averaging the triplicate samples used. The percent degradation presented in the plot accounts for all the losses reported in the control samples, which were less than 10%, a reasonable loss for complex systems such as this one. As can be seen in the figure, the percent biodegradation was 23% after the first day of incubation in the microcosm. Thereafter the biodegradation increased slowly but steadily as indicated the positive slope in the plot to about 55% by the end of a six-week incubation period. The positive slope of the biodegradation curve indicates that neither oxygen nor nutrients were limiting microbial growth. It is possible then that the degradation of 1,3-DCB will proceed, albeit at a very slow rate, if the experiments were continued beyond the six-week period. It was noted in our earlier studies that about 60% (F=0.6 from Eq. 1) of the sorbed contaminant, 1,3-DCB, is reversibly bound and the rest being in the non-labile phase or tightly bound to soil [7, 9, 10]. It appears from the figure that bacteria were able to degrade a significant fraction of the readily available fraction of 1,3-DCB.

As would be expected, the rate of biodegradation of DCB was much slower in the soil phase than that in the aqueous phase. For example, nearly 80% of 1,3-DCB was biodegraded within three days of incubation and nearly complete degradation in seven days for the aqueous phase (soil free) compared to about

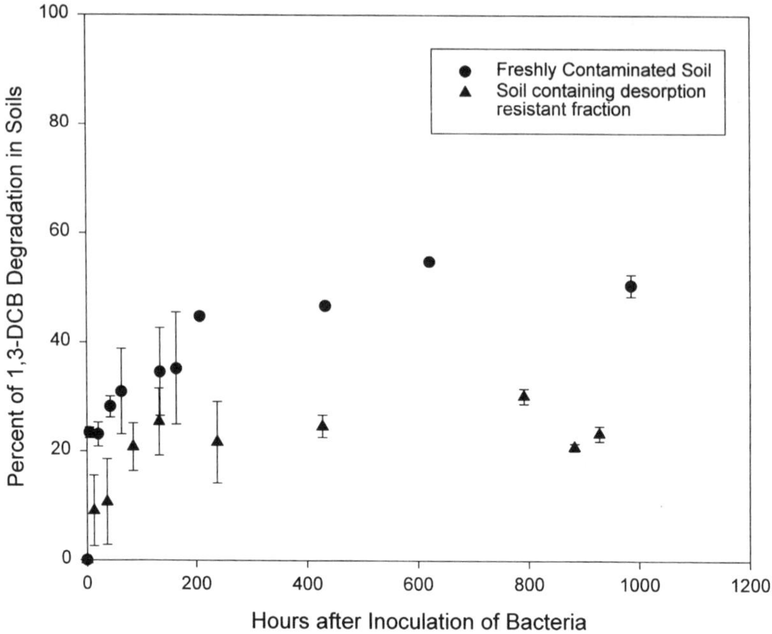

Fig. 7 Biodegradation of 1,3-DCB from freshly contaminated soil and soil containing only irreversibly bound (non-labile) 1,3-DCB. Percent degradation was calculated after accounting for the losses in the controls

55% degradation from freshly contaminated soil. This is in agreement with the findings from the literature [20, 23]. It is also reported that microorganisms utilize the substrate (contaminant) from the aqueous phase rather than metabolize the contaminant directly sorbed to the soil. Hence mass transfer of 1,3-DCB from soil surface to the aqueous phase could play a significant role and potentially limit biodegradation [23]. However, for the readily desorbing fraction, mass transfer (desorption) does not seem to limit biodegradation. However, this will not be the case for experiments with the soil containing desorption resistant fraction as discussed in the following section.

2.5.2 Desorption-Resistant Fraction

A known amount of the soil that was contaminated and subjected to sequential desorption, containing only the desorption-resistant fraction of DCB, was added to the microcosms, which were prepared as described earlier. The biodegradation of 1,3-DCB was monitored with time over a six-week incubation period. As noted earlier, the initial soil concentration was about 1650 mg/kg soil compared to 18,000 mg/kg soil for freshly contaminated soil experiments. The percent biodegradation was calculated as above after accounting for the losses from the control samples. Figure 7 displays percent biodegradation as a function of incubation time for the soil containing the tightly bound fraction of 1,3-DCB. As shown in the figure, degradation from the soil containing non-labile phase DCB was significantly lower than the labile phase degradation. For example, 1,3-DCB degradation was about 22% during the first week with only a total of 33% during the total six-week incubation period. It should be noted that triplicate samples were sacrificed at each sampling period and slight variations in data were noted among the triplicates, which is a fairly common occurrence. Also, it is important to note that the soil used in the experiments is assumed to have only a desorption resistant fraction. While the soil was subjected to five sequential desorption steps before using in the experiment, it is possible that it may still contain a fraction which is readily available for microbes during the initial incubation period as indicated by the initial high rate of biodegradation. However, after the first week, the 1,3-DCB biodegradation rate was very small, as one would expect. Desorption from the tightly bound non-labile phase is thus limiting the availability of the contaminant for biodegradation. Microorganisms metabolize the substrate present in the aqueous phase rather than direct metabolism from the soil phase. Though there have been some reports that sorbed substrate may be directly available for degradation by attached cells either by direct partitioning to the cell membrane or via degradation by extracellular enzymes [40], there is still disagreement among the researchers on this point.

Figure 7 also provides a comparison between freshly contaminated soils and soil with a desorption resistant fraction. As evident from the figure, the degradation in soil containing the tightly bound contaminant was 30% compared to about 55% for freshly contaminated soil. A number of investigators have shown that the desorption resistant fraction in the soil is unavailable for microorganisms even under optimal conditions for biodegradation [20, 22–25]. The tightly bound contaminant could be present in the soil micropores or chemically bound to soil humic matter and thus microorganisms are not be able to access this fraction. This is supported by data from Robinson et al. [23] on toluene biodegradation and Ogram et al. [21] who reported that microorganisms could not utilize the sorbed phase contaminant. Our microcosm batch studies showed that microorganisms readily degraded the easily extractable or readily desorbed fraction of the sorbed DCB and that desorption mass transfer is limiting the growth of the microorganisms and thus the rate of biodegradation for the desorption resistant fraction.

3
Implications for Remediation at the PPI Sites

The observed hysteresis (the bi-phasic nature) in the chemical sorption/desorption process has profound implications as far as the current remedy for the PPI site is concerned. In order to illustrate this aspect, we need to use an appropriate model for desorption from the soils and dissolution in the groundwater. A model was developed by Thibodeaux [41] to estimate the concentration in groundwater from an aquifer contaminated with a hydrophobic organic compound. The volume of the contaminated zone is divided into n sub-volumes of equal size, and the concentration of water exiting this contaminated zone (withdrawal well) is obtained. The exit concentration of the contaminant, C_w in the porewater relative to the initial concentration C_w^o is then given by

$$\frac{C_w}{C_w^o} = \exp\left(-\frac{nt}{\tau_c}\right)\left[1 + \frac{nt}{\tau_c} + \ldots + \frac{\left[\frac{nt}{\tau_c}\right]^{(n-1)}}{(n-1)!}\right] \tag{6}$$

In the above equation t is the time (years) and τ_c is the chemical residence time (years) given by the following equation:

$$\tau_c = [\varepsilon + K_{sw}(1-\varepsilon)\rho_b]\frac{L}{V_x} = R_F \cdot \frac{L}{x} \tag{7}$$

For Case I we choose $K_{sw}=K_{sw}^{des,1}$, whereas for Case II we choose $K_{sw}=K_{sw}^{des,2}$. The corresponding retardation factors are $R_{F1}=750$ and $R_{F2}=11{,}250$ for the two cases. For illustrative purposes, we choose a 10-m zone length with a soil porosity (ε) of 0.25. An 'n' value of three is chosen in keeping with the fact that the changes in concentrations predicted are only marginally affected for values of 'n' greater than three. L is the length of the zone (10 m) and V_x is the groundwater velocity, which is taken to be approximately 3.3 m/year. For Case I, the initial concentration is given by $\rho_b W^o/R_{F1}$=2120 μg/l as in Fig. 4b. For Case II, the initial aqueous concentration is re-scaled to 160 μg/l using the sediment concentration for the irreversible fraction (1043 μg/g) and R_{F2}. Figure 8 shows model results as the water concentration at the exit (withdrawal well) of the 10 m zone vs time for HCBD at the PPI site. Under the assumptions made above, the exit concentration in water from the zone decreased from 2280 μg/l initially to 2178 μg/l after 100 years, 648 μg/l after 500 years, and 173 μg/l after 1000 years. Even after 5000 years, the concentration in the water was small, but measurable (17 μg/l) and was determined by the gradual leaching of the tightly bound fraction. Note that the national recommended water quality criterion for HCBD is 0.44 μg/l, which would be reached in about 9600 years. The area under the curve represents the mass removed and in 1000 years was only 17% of the total. However, if the partitioning was entirely reversible, the predicted aqueous concentration decreased to only 1,555 μg/l after 1000 years and the mass recovery was 41%. Due to the continuous removal of the reversibly sorbed HCBD, the predicted concentration in the aqueous phase after 5000 years was only 6.8 μg/l in this case. As expected, the concentration in the withdrawal well is predicted to be approximately three times as large over the long term when the process is only partly reversible than when the adsorption is completely reversible. Clearly, the total mass recovery of the contaminant (i.e., HCBD) from the aquifer by pump and treat is expected to be very small. A large fraction of

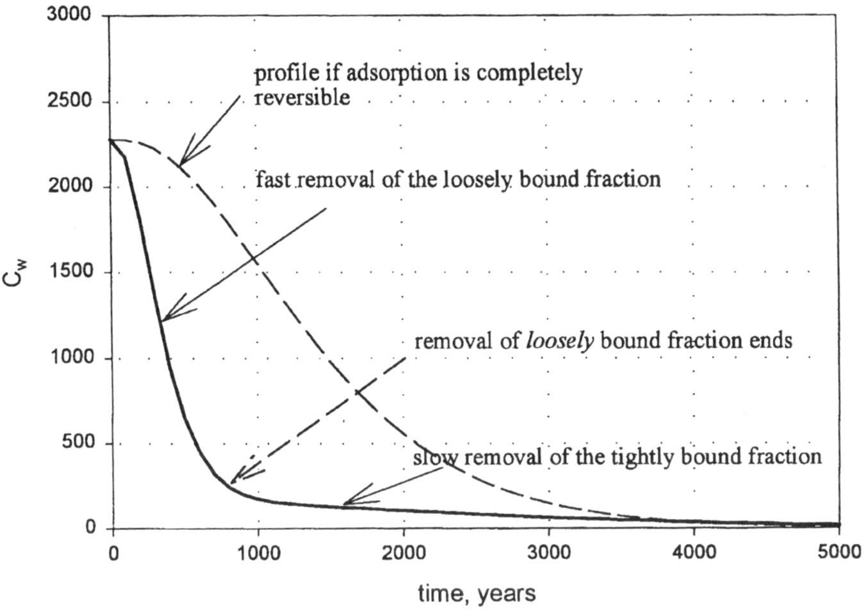

Fig. 8 Predicted HCBD in a withdrawal well during pump and treat remediation of a 10-m contaminated zone at the PPI site

the mass (67%) is tightly bound to the soil and will take several thousand years to be removed by conventional groundwater removal technology.

There are several important consequences of the above findings. First, it is clear that satisfactory removal of residual soil-sorbed HCBD (i.e., the fraction left after the free phase removal) by conventional pump-and-treat is difficult and requires an inordinate and unacceptable time frame. In other words, much of the HCBD is "irreversibly" bound to the soil. If recovery is the only remedy, other enhanced removal schemes should be considered. Second, since most of the HCBD is bound to the soil and leaches only slowly, and since microbes capable of consuming HCBD are known, monitored natural attenuation or enhanced bioremediation are better options for site remediation if removal is not required. This would, of course, require a continuous monitoring of the plume so that no off-site migration of the contaminants occurs during the site cleanup, which is required in MNA projects. Third, since HCBD predominantly remains bound to soil particles, its movement off-site with the groundwater is likely not to be significant, and justifies its consideration as a good candidate compound for MNA.

The calculations made above are approximate and neglect many other factors such as the site heterogeneity, biodegradation, and contaminant concentration variations. However, the basic conclusions, with respect to the difficulty in extracting HCBD and benefits it illustrates in MNA, should remain the same even if a more complete and detailed model such as MODFLOW is considered.

Experimental results and basic modeling of transport suggest that at most only 60% of the compound is labile and participates in the reversible sorption equilibrium. The half-life for desorption of this fraction is of the order of a few hours (2–24 h) for TCE (from Lee et al. [9]), HCBD and larger for DCB. The labile fraction becomes smaller as the contaminant ages within the soil. Our earlier work [8]showed that even when water in contact with the soil is replaced with fresh water at every step after 24 h of equilibration only a small percentage of the material is recovered from the contaminated soil. Hence, sequential desorption also is incapable of removing the non-labile fraction from the soil. The projections from the above models based on the 72-h equilibrium data only pertain to the labile fraction of pollutant. Hence, the long term predictions using the batch K_d values will significantly over-predict movement of contaminants that have been in contact with the soil for decades at the PPI site. On the other hand, we conclude that a significant portion of the contaminant in the aged site soil is inaccessible to water in a conventional recovery pumping scheme and hence remain within the soil to pose no significant threat of migration away from the site. Moreover, the slowly released fraction can probably be managed by natural assimilative capacity of the soil, given that biodegradation and sorption are common at the site(s). Thus, the desorption resistant concentration may be considered an environmentally acceptable end point (EAE) in the soil that can be managed by MNA procedures currently in practice.

Monitored Natural Attenuation, as mentioned previously, is currently the remedy of choice for Brooklawn and Scenic sites at PPI. Implementation of MNA is in-part due to results of research presented in this chapter as well as work of others at LSU, NPC, and associated parties to the case. If one looks at the site history, the remedy has changed over the last decade from very active (excavation) through active (hydraulic containment and recovery) to passive methods (MNA). While active remediation did reduce volumes of waste at the site prior to capping and groundwater pumping, emissions were a problem. In addition to implementation of MNA, capping and monitored pumping is planned for certain areas outside of Brooklawn and Scenic in addition to MNA, but these areas are relatively small. Thus, pending results of monitoring of contamination at the site and completion of detailed modeling beyond MODFLOW (using MT3D® and STOMP© (Battelle) for groundwater and NAPL, respectively), MNA is the solution at this time. Our work has provided several conclusions directing the remedy to MNA. First, simple models employed herein, as well as more sophisticated ones, show that it will take many years to obtain remediation of the contamination down to target levels by current technology. Not only is this timeframe unacceptable (thousands of years) but additional work by NPC has shown that continued active pumping of groundwater: (1) increases the volume of waste to handle, and (2) may actually spread the DNAPL contamination beyond its current footprint. Thus the active process is actually working against natural processes with basis in our findings, i.e., the easily desorbed fraction biodegrades at high rates in situ and the slowly desorbed fraction binds contaminants to the point of rate limiting degradation. In other words, detailed MNA models [42] for NAPL and groundwater show that pumping has little effect on the NAPL at the sites and conditions appear to be favorable for significant degradation (and sorption in any movement) of contaminants in the groundwater. Continued modeling efforts and long term monitoring will determine if direction from this work and the remedy are environmentally acceptable over the long term. Monitoring in the short term (1990s and last two years) certainly points in this direction incorporating risk assessment as well as resulting in significant reduction in the cost of cleanup. It is anticipated that if the remedy is found effective in the short term, and modeling indicates MNA is acceptable, that this highly ranked NPL site will be de-listed, with long term monitoring continuing to ensure protection of human health and the environment. It is also possible that some day this site can be recycled (used again) for normal industrial activities like the neighboring areas where several industrial plants are located.

References

1. Environmental Protection Agency (2002) http://www.epa.gov/earth1r6/6sf-1a.htm
2. Mackay DM, Cherry JA (1989) Groundwater contamination, pump and treat. Environ Sci Technol 23(6):630–634
3. Hunt JR, Sitar N (1988) Non-aqueous phase liquid transport and clean-up. 1. Analysis of mechanisms. Water Resour Res 24(8):1247–1258
4. NPC Services (1995) Remedial planning activities (RPA) for the PPI site, vol II. Baton Rouge, LA
5. Kommalapati RR (1995) Remediation of contaminated soils using a plant-based surfactant. Ph. D Dissertation, Louisiana State University, Baton Rouge, LA
6. Kommalapati RR, Valsaraj KT, Constant WD (2000) Soil-water partition coefficients, adsorption/desorption hysteresis, desorption kinetics and bioavailability of chlorinated organic compounds at the PPI site. Submitted to Hazardous Substance Research Center, Louisiana State University, Baton Rouge, LA
7. Valsaraj KT, Kommalapati RR, Robertson ED, Constant WD (1999) Partition constants and adsorption/desorption hysteresis for volatile organic compounds on soil from a Louisiana Superfund site. Environ Monit Assess 58:225–241
8. Kommalapati RR, Valsaraj KT, Constant WD (2002) Soil-water partitioning and desorption hysteresis of volatile organic compounds from a Louisiana Superfund site soil. Environ Monit Assess 73(3):275–290
9. Lee S, Kommalapati RR, Valsaraj KT, Pardue JH, Constant WD (2002) Rate-limited desorption of volatile organic compounds from soils and implications for the remediation of a Louisiana Superfund site. Environ Monit Assess 75(1):93–111
10. Lee S, Kommalapati RR, Valsaraj KT, Pardue JH, Constant WD (2002) Bioavailability of reversibly sorbed and desorption resistant 1,3-dichlorobenzene from a Louisiana Superfund site soil. Water Air Soil Pollut (in review)
11. Kan AT, Fu G, Hunter MA, Tomson MB (1997) Irreversible adsorption of naphthalene and tetrachlorobiphenyl to Lula and surrogate sediments. Environ Sci Technol 31:2176–2181
12. Mackay AA, Chin Y, Macfarlane JK, Gschwend PM (1996) Laboratory assessment of BTEX soil flushing. Environ Sci Technol 30(11):3223–3231

13. Pignatello JJ, Xing B (1996) Mechanisms of slow sorption of organic chemicals to natural particles. Environ Sci Technol 30:1–11
14. Alexander M (1994) Biodegradation and bioremediation. Academic Press, New York
15. Di Toro DM, Horzempa LM (1982) Reversible and resistant components of PCB adsorption-desorption isotherms. Environ Sci Technol 16:594–602
16. Karickoff SW, Morris KW (1985) Impact of tubificid oligochaetes on pollutant transport in bottom sediments. Environ Toxicol Chem 9:1107–1115
17. Opdyke DR, Loehr RC (1999) Determination of chemical release rates from soils: experimental design. Environ Sci Technol 33:1193–1199
18. Linz DG, Nakles DV (1997) Environly acceptable endpoints in soil: risk based approach to contaminated site management based on availability of chemicals in soil. American Academy of Environ Engineers, New York, NY
19. Bosma TNP, Middeldorp PJM, Schraa G, Zehnder AJB (1997) Mass transfer limitation of biotransformation: quantifying bioavailability. Environ Sci Technol 31(1):248–252
20. Zhang W, Bouwer EJ (1997) Biodegradation of benzene, toluene and naphthalene in soil-water slurry microcosms. Biodegradation 8:167–175
21. Ogram AV, Jessup RE, Ou LT, Rao PSC (1985) Effects of sorption on biological degradation rates of 2,4-dichlorophenoxy acetic acid in soils. Appl Environ Microbiol 49:582–587
22. Al-Bashir B, Hawari J, Samson R, Leduc R (1994) Behavior of nitrogen-substituted naphthalenes in flooded soil. Part II. Effect of bioavailability of biodegradation kinetics. Water Res 28(8):1827–1833
23. Robinson KG, Farmer WS, Novak JT (1990) Availability of sorbed toluene in soils for biodegradation by acclimated bacteria. Water Res 24:345–350
24. Steinberg SM, Pignatello JJ, Sawhney BL (1987) Persistence of 1,2-dibromoethane in soils: entrapment in intraparticle micropores. Environ Sci Technol 21:1201–1208
25. Pignatello JJ (1989) Sorption dynamics of organic compounds in soils and sediments. In: Sawhney BL, Brown K (eds) Reactions and movement of organic chemicals in soils. Soil Science Society of America and American Society of Agronomy, pp 31–80
26. Valsaraj KT (1995) Elements of environmental engineering. CRC Press, Lewis Publishers, Boca Raton, Fl
27. Schwarzenbach RP, Gschwend PM, Imboden DM (1993) Environmental organic chemistry. Wiley, New York, NY
28. Carroll KM, Harkness MR, Bracco AA, Balcareel RB (1994) Application of a permeant/polymer diffusional model to the desorption of polychlorinated biphenyls from Hudson River sediments. Environ Sci Technol 28:253–258
29. Hatzinger AB, Alexander M (1995) Effects of aging of chemicals in soils on their biodegradability and extractability. Environ Sci Technol 29:537–545
30. Kan AT, Fu G, Hunter M, Chen W, Ward CH, Tomson MB (1998) Irreversible sorption of neutral hydrocarbons to sediments: experimental observations and model predictions. Environ Sci Technol 32:892–902
31. Gess P, Pavlostathis SG (1997) Desorption of chlorinated organic compounds from a contaminated estuarine sediment. Environ Toxicol Chem 16(8):1598–1605
32. Pavlostathis SG, Mathavan GN (1992) Desorption kinetics of selected volatile organic compounds from field contaminated soils. Environ Sci Technol 26:532–538
33. Locke MA (1992) Sorption-desorption kinetics of Alachlor soil from two soybean tillage systems. J Environ Qual 21:558–566
34. Xing B, Pignatello JJ (1996) Time-dependent isotherm of organic compounds in soil organic matter: implications for sorption mechanism. Environ Toxicol Chem 15(8): 1282–1288
35. Montgomery JM (1997) Groundwater chemicals desk reference. CRC Press, Boca Raton, FL

36. Pavslostathis SG, Jaglal K (1991) Desorptive behavior of trichloroethylene in contaminated soil. Environ Sci Technol 25(2):274–279
37. Pignatello JJ, Ferrandino FJ, Huang LQ (1993) Elution of aged and freshly added herbicides from a soil. Environ Sci Technol 27:1563–1571
38. Connaughton DF, Stedinger JR, Lion LW, Shuler ML (1993) Description of time-varying desorption kinetics: release of naphthalene from contaminated soil. Environ Sci Technol 27:2397–2403
39. Scribner SL, Benzing TR, Sun S, Boyd SA (1992) Desorption and bioavailability of aged Simazine residues in soil from a continuous corn field. J Water Qual 21:115–120
40. Park JH, Zhao X, Voice TC (2001) Biodegradation of non-desorbable naphthalene in soils. Environ Sci Technol 35(13):2734–2740
41. Thibodeaux LJ (1996) Environmental chemodynamics. Wiley, New York, NY
42. NPC Services (2002) Petro-Processors Site Update to EPA, LDEQ and LSU, Presentation with Battelle Inc., May 7, 2002, Dallas, TX

Handb Environ Chem Vol. 5, Part F, Vol. 3 (2005): 159–189
DOI 10.1007/b11732

Solidification/Stabilization Technologies for the Prevention of Surface and Ground Water Pollution from Hazardous Wastes

Marina R. Ilic[1] (✉) · Predrag S. Polic†[2]

[1] Institute for General and Physical Chemistry, Studentski trg 12/V, 11000 Belgrade, Serbia and Montenegro
marinai@Eunet.yu

[2] Department of Chemistry, University of Belgrade, Studentski trg 14, 11000 Belgrade, Serbia and Montenegro

1 Introduction . . . 161

2 Waste Classification . . . 161
2.1 Classification System . . . 161
2.1.1 Hazardous Wastes from Non-Specific Sources . . . 162
2.1.2 Hazardous Wastes from Specific Sources . . . 162
2.1.3 Commercial Chemical Products . . . 162
2.2 Toxicity, Corrosivity, Ignitability, and Reactivity . . . 162
2.2.1 Toxicity . . . 163
2.2.2 Corrosivity . . . 163
2.2.3 Ignitability . . . 163
2.2.4 Reactivity . . . 163

3 Technologies for Hazardous Waste Treatment . . . 164
3.1 Solidification/Stabilization . . . 164
3.1.1 Cement-Based Stabilization/Solidification . . . 165
3.1.2 Lime-Based Stabilization/Solidification . . . 166
3.1.3 Pozzollanic Techniques . . . 166
3.1.4 Thermoplastic Stabilization . . . 166
3.1.5 Solidification/Stabilization with Organic Polymers . . . 166
3.1.6 Encapsulation . . . 167
3.1.7 Solidification/Stabilization by Self-Cementation . . . 167
3.2 Chemical Treatment Techniques . . . 167
3.2.1 Oxidation/Reduction . . . 168
3.2.2 Solvent Recovery/Extraction . . . 169
3.3 Physico-Chemical Treatment Techniques . . . 169
3.3.1 Adsorption on Activated Carbon . . . 170
3.3.2 Distillation . . . 170
3.3.3 Exchange of Ions . . . 170
3.3.4 Reverse Osmosis . . . 171
3.4 Thermal Treatment Techniques . . . 171
3.4.1 Thermal Desorption/Destruction . . . 172
3.4.2 Vitrification . . . 172
3.4.3 Plasma Heating . . . 172
3.4.4 Microwave Heating . . . 172
3.4.5 Incineration . . . 173

4 **Mechanisms of Leaching Processes** . . . 173
4.1 Fundamentals of Leaching . . . 173
4.2 Commonly Used Leaching Tests on Solidified/Stabilized Waste Products . . . 174
4.3 Leaching Studies . . . 174
4.3.1 Column . . . 174
4.3.2 Batch . . . 174
4.3.3 Cascade . . . 175
4.3.4 TCLP . . . 175
4.3.5 GANC . . . 175
4.4 Parameters of Leaching Processes . . . 175
4.5 Physico-Chemical Mechanisms of Mass Transfer . . . 176
4.6 Data Interpretation of Leaching Tests . . . 176
4.7 Environmental Impact Assessment on Surface and Ground Waters . . . 176

5 **A Case Study: Solidification/Stabilization of Lead and Cadmium Ions in the Cement Matrix** . . . 177
5.1 Overview . . . 177
5.2 Experimental Work . . . 177
5.3 Results and Discussion . . . 178
5.4 Summary of Findings . . . 186

6 **Conclusions and Recommendations** . . . 186

References . . . 187

Abstract Solidification/stabilization treatment processes immobilize hazardous constituents in the waste by changing these constituents into immobile (insoluble) forms, binding them in an immobile matrix, and/or binding them in a matrix which minimizes the material surface exposed to weathering and leaching. Solidification/stabilization treatment processes can include aluminum silicate and cement-based fixation, pozzolanic-based fixation, or vitrification.

The process of solidification/stabilization is a widely accepted treatment/disposal process for a broad range of wastes, particularly those classified as toxic or hazardous, which are not suited for normal methods of disposal and where special treatment is necessitated. Portland cement is a material found to be most useful for solidification/stabilization purposes due to its ability for heavy metals fixation and immobilization. For the immobilization of waste containing high concentrations of heavy metals, as in the case of the galvanization process, solidification is a very acceptable treatment. This is also consistent with the recent EU directions, which refereed to the solid waste management. Solidification/stabilization processes will play a more important role since, in the near future, only inert or stabilized wastes should be landfilled. Solidification/stabilization process means binding the hazardous material in the hydraulic binders for safe landfilling or use in civil engineering purposes. Various types of cement and pozzolanas (e.g., coal burning fly ash, lime, blast-furnace slag and similar materials) are mostly used as the stabilizing matrix. Those stabilization techniques are used for the immobilization of inorganic or organic waste. The end product of the treatment, usually after sufficient curing, is solid monolithic material which, depending on characteristics of leaching, can be usefully applied or disposed of in a safe way.

Keywords Solidification/stabilization · Hazardous material · Leaching processes

List of Abbreviations

US EPA	United States Environmental Protection Agency
TCLP	Toxicity characteristic leaching procedure
ANC	Acid neutralizing capacity procedure
PCB	Polychlorinated biphenyls
NEN	Dutch standard
GANC	Generalized acid neutralization capacity
JUS	Yugoslav standard

1 Introduction

Stabilization techniques are used for the immobilization of inorganic or organic waste. Stabilization procedure is relatively simple which represents one advantage more of such way of waste management. Stabilization refers to techniques that chemically reduce the hazard potential of a waste by converting the contaminants into less soluble, mobile, or toxic forms. The physical nature and handling characteristics of the waste are not necessarily changed by stabilization. Stabilization, or fixation, is safe way for treatment of sludge, certain materials, liquids and radioactive waste before final disposal. Benefits of stabilizing hazardous wastes include: (a) improving the physical properties of the wastes for easier handling; (b) safer transport and easier burial; (c) detoxifying various wastes for the protection of workers; (d) preventing environment pollution caused by leaching and evaporation of hazardous constituents; and (e) recycling into construction material.

Solidification/stabilization is a promising treatment technology for containing and immobilizing dredged material contaminants within a disposal site. While solidification/stabilization is not a solution to every disposal problem, the technology offers improved physical characteristics that reduce the accessibility of water to contaminated solids and reduced leachability for many contaminants. Leaching tests must be able to determine the: (a) compounds that can be released from the waste; (b) maximum concentration of these compounds in the leachate; (c) quantities released per unit mass of waste; (d) release rate of these compounds; and (e) effects of a co-disposal of the waste.

2 Waste Classification

2.1 Classification System

Data about appearance of waste chemicals in industrial and municipal wastes is the primary criteria which determines the significance of waste chemicals for health and environment. Two secondary criterions of equal significance are:

(a) the possibility of chemicals to be found and to transport in nature, and (b) the possibility of chemicals to lead to health deviations. In terms of categorization of hazardous waste sources the US EPA has grouped hazardous waste in to three groups [1], as follows.

2.1.1 Hazardous Wastes from Non-Specific Sources

Generic wastes produced by manufacturing and industrial processes are included in the list of hazardous wastes from non-specific sources. Spent halogenated solvents, bottom sludge from electroplating operations, wastes from various chemicals manufacturing, and the like are examples of hazardous wastes from non-specific sources. These are designated by prefix F [2–4].

2.1.2 Hazardous Wastes from Specific Sources

This list includes sludge, still bottoms, wastewaters, spent catalysts, and residues. These are generated by specific industrial processes, such as wood preserving, petroleum refining, and organic chemical manufacturing. These are designated by prefix K [5, 6].

2.1.3 Commercial Chemical Products

This list comprises chemicals like chloroform, creosote, acids, pesticides, and numerous other commercial chemicals. These are designated by prefix P and U [7, 8].

2.2 Toxicity, Corrosivity, Ignitability, and Reactivity

The final step to determine whether a waste is hazardous is to evaluate the waste against four hazardous characteristics. These characteristics are toxicity, corrosivity, ignitability and reactivity. A toxic waste is harmful to human or other forms of life, it may be carcinogenic, mutagenic, or teratogenic, and it may cause death. A reactive hazardous waste may be explosive, may form a dangerous mixture with water, or may undergo chemical change without addition of another substance to form a toxic product. An ignitable waste may be a fire hazard. A waste is considered corrosive if it is aqueous and has a pH of 2 or less or 12.5 or more [9]. The following is a summary.

2.2.1 Toxicity

The fourth characteristic which could make a waste a hazardous waste is toxicity. To determine if a waste is a toxic hazardous waste, a representative sample of the material must be subjected to a test conducted in a certified laboratory. The test procedure used from the inception of the hazardous waste management program was the EP-Toxicity test. The new test procedure is the Toxicity Characteristic Leaching Procedure (TCLP). The TCLP is more aggressive than was the formerly used EP-Toxicity test for certain chemicals. Many industrial wastes which previously were not hazardous wastes may be considered hazardous waste when tested using the new procedure.

2.2.2 Corrosivity

A liquid waste which has a pH of less than or equal to 2 or greater than or equal to 12.5 is considered to be a corrosive hazardous waste. Sodium hydroxide, a caustic solution with a high pH, is often used to clean or degrease metal parts. Hydrochloric acid, a solution with a low pH, is used by many industries to clean metal parts prior to painting. When these caustic or acid solutions become contaminated and must be disposed, the waste would be a corrosive hazardous waste.

2.2.3 Ignitability

A waste is an ignitable hazardous waste if it has a flash point of less than 60 °C, readily causes fires and burns so vigorously as to create a hazard. A simple method of determining the flash point of a waste is to review the material safety data sheet, which can be obtained from the manufacturer or distributor of the material. Naphtha, lacquer thinner, epoxy resins, adhesives, and oil based paints are all examples of ignitable hazardous wastes.

2.2.4 Reactivity

A material is considered to be a reactive hazardous waste if it is normally unstable, reacts violently with water, generates toxic gases when exposed to water or corrosive materials, or if it is capable of detonation or explosion when exposed to heat or a flame. Examples of reactive wastes would be waste gunpowder, sodium metal or wastes containing cyanides or sulfides [10].

3 Technologies for Hazardous Waste Treatment

3.1 Solidification/Stabilization

Solidification/stabilization treatment processes immobilize hazardous constituents in the waste by changing the constituents into immobile (insoluble) forms, binding them in an immobile, insoluble matrix, and/or binding them in a matrix which minimizes the material surface exposed to weathering and leaching. Solidification/stabilization treatment processes can include aluminum silicate and cement-based fixation, pozzolanic-based fixation, or vitrification. Once they have been separated from aqueous solution and concentrated in ash or sludge, the hazardous constituents must be bound up in stable compounds. Reduced leachability is accomplished by the formation of a lattice structure and/or chemical bonds that bind the hazardous constituent and thereby limit the amount of constituent that can be leached when water or a mild acid solution comes into contact with the waste matrix. There are two principal solidification/stabilization processes: cement based and lime based. The cement or lime additive is mixed with the ash or sludge and water. It is then allowed to cure to form a solid. In both techniques the stabilizing agent may be modified by other additives such as silicates. In general, this technology is applicable to wastes containing metals with little or no organic contamination, oil, or grease [11–15].

Solidification and stabilization are generic names applied to a wide range of discrete technologies that are closely related in that both use chemical and/or physical processes to reduce potential adverse impacts on the environment from the disposal of radioactive, hazardous, and mixed wastes [16–18].

Table 1 Methods and applications of stabilization

Stabilization method	Application
Solidification with cement	Sludges, contaminated soil, various metal salts, low-level radioactive waste
Solidification by lime-based processes	Flue gas desulfurization wastes, other inorganic wastes
Solidification with thermoplastic materials	Radioactive wastes
Solidification with organic polymers	Sludges, radioactive sludge
Encapsulation	Sludges, liquids, particulate matter
Solidification by self-cementation	Flue gas desulfurization wastes, other wastes with large proportions of calcium sulfate or calcium sulfite
Vitrification	Extremely hazardous wastes, radioactive wastes

Solidification can be accomplished by a chemical reaction between the waste and solidifying reagents or by mechanical processes. Contaminant migration is often restricted by decreasing the surface area exposed to leaching and/or by coating the wastes with low-permeability materials. The combined process of solidification/stabilization mixes wastes, soils, and sludges with treatment agents to immobilize, both physically and chemically, the hazardous constituents in those substances. The technologies are not regarded as destructive techniques; rather, they eliminate or impede the mobility of contaminants [19–22]. Many methods of stabilization can be classified according to the seven categories presented in Table 1.

3.1.1 Cement-Based Stabilization/Solidification

Solidification with cement generally is accomplished with a Portland cement and other additives. The quantity of cement can be varied according to the amount of moisture in the waste. Heavy metal cations in the waste form insoluble carbonates and hydroxides at the high pH of the mixture. The surface of the hardened mass can be coated with asphalt or other material to reduce leaching of hazardous components. If the waste is mixed with anhydrous cement and water there is the possibility of ions incorporation in the cement structure during the hydrolysis process. Heavy metal ions could bind with the cement by the process of chemisorption, precipitation, surface adsorption,

Table 2 Waste components for which the cement solidification have restricted application

Waste component	Defect
Highly soluble metals or salts which contain As, Ni, soluble salts, Mn, Sn, Zn, Cu, Pb, Na	Influence on cement setting and could be leached easily
Highly toxic anions, borates	Much easier leaching than cations
High content of cement setting retarders like sulfates, halides	Lower strengths of the final product
Inflammable/explosive materials	Safety danger
Biologically active compounds such as insecticides, pesticides	Can be leached easier than final products
Compounds which prevent cement setting like sugar, lignine sludge	Lower strengths of the final product
Production of toxic/hazardous gases in the contact with water like metal carbides	Health danger
Compounds which does not easily refrain in cement like phenols	Easy leaching

"capturing" inside the matrix, chemical incorporation or with the combination of mentioned possibilities [23–28].

It is also known that small quantities of some compound could significantly reduce strengths and deteriorate the properties of cement solidified waste. That could be in the form of setting reduction, which could lead to the material disintegration and leaching. Examples of some waste components, which could give opposite effect to the cement stabilization, are given in Table 2.

3.1.2 Lime-Based Stabilization/Solidification

Lime/fly ash pozzolanic processes combine the properties of lime and fly ash to produce low-strength cementation. Kiln dust processes involve the addition of kiln dust to eliminate free liquids and usually form a low-strength solid. Lime-based processes for solidification use reactions of lime with water and pozzolanic (siliceous) materials, such as fly ash or dust from cement kilns, to form concrete, called a pozzolanic concrete. Wastes of desulfurization of gases and other inorganic wastes can be immobilized by this method.

3.1.3 Pozzollanic Techniques

Various types of cement and pozzolanas (e.g., coal burning fly ash, lime, blast-furnace slag and similar materials) are mostly used as the stabilizing matrix. That stabilization technique is used for the immobilization of inorganic or organic waste.

3.1.4 Thermoplastic Stabilization

Thermoplastic materials for solidification, such as bitumen, polyethylene, and paraffin, are mixed with dried wastes at elevated temperatures. The mixtures solidify when they cool. The hardened mixtures may be placed into containers prior to disposal. One group of materials not suitable for this process, however, includes organic wastes that can dissolve the thermoplastic matrix and thus prevent solidification. Chlorates, perchlorates, and nitrates in high concentrations can deteriorate bitumen. Radioactive wastes can be immobilized by this method.

3.1.5 Solidification/Stabilization with Organic Polymers

Methods of solidification with organic polymers involve the addition and thorough mixing of monomers, such as urea and formaldehyde, with the wastes. A polymerization catalyst can be added. The mixtures can be placed in containers prior to disposal. Sludges and radioactive wastes can be immobilized by this method.

3.1.6 Encapsulation

Solidification refers to techniques that encapsulate the waste, forming a solid material, and does not necessarily involve a chemical interaction between the contaminants and the solidifying additives. The product of solidification, often known as the waste form, may be a monolithic block, a clay-like material, a granular particulate, or some other physical form commonly considered "solid." Solidification as applied to fine waste particles, typically 2 mm or less, is termed microencapsulation and that which applies to a large block or container of wastes is termed macroencapsulation [29].

The method of encapsulation involves covering of wastes with inert and impervious coating or jackets, which will prevent leaching. Metal drums of 200 l capacity, which can become corroded by hazardous wastes, can be encapsulated with polyethylene jackets.

3.1.7 Solidification/Stabilization by Self-Cementation

Methods of solidification by self-cementation are applicable to wastes that contain large amounts of calcium sulfate or calcium sulfite. A small quantity of the waste is dried, calcined, and added to the rest of the waste with other additives to form a hardened mass [30].

3.2 Chemical Treatment Techniques

Chemical stabilization is the alteration of the chemical form of the contaminants to make them resistant to aqueous leaching. Solidification/stabilization processes are formulated to minimize the solubility of metals by controlling pH and alkalinity. Entrapment or microencapsulation may immobilize anions, which are more difficult to bind in insoluble compounds. Chemical stabilization of organic compounds may be possible, but the mechanisms involved are poorly understood [22].

Deactivation by chemical conversion is a process by which a waste or constituent of the waste is transformed to at least one substance that is less hazardous and chemically different from the original material. Generally, methods of removing hazardous constituents can be used for concentrating the hazardous constituents and reducing the volume of hazardous waste.

Many wastes can be chemically converted to less hazardous products and organic compounds (Table 3).

The spectrum of chemical methods includes complexation, neutralization, oxidation, precipitation and reduction. An optimum method would be fast,

Table 3 Chemical conversion processes for hazardous waste

Process	Waste type
Wet oxidation	Variety of organic compounds
Ozonation	Phenols, cyanide, and organic lead com pounds in wastewaters, compounds in air
Molten-salt combustion	Variety of organic compounds
Electrochemical oxidation	Cyanate, thiocyanate, acetate, phenols, cresols
Treatment with formaldehyde	Chromium (VI), cyanide ion, metal-cyanide complexes
Catalytic reduction with metal powder	Chlorinated organic compounds in wastewaters
Catalytic hydrogenation	Polychlorinated hydrocarbons
Dechlorination	Polychlorinated organic compounds
Destruction by microwave plasma	Pesticides, polychlorinated biphenyls
Hydrolysis	Organophosphorus pesticides, carbamate pesticides
Neutralization	Strong acid, alkalies

quantitative, inexpensive, and leave no residual reagent, which itself would be a pollution problem.

3.2.1 Oxidation/Reduction

Hazardous waste could be successfully and mostly cheaply eliminated with those methods. Strong oxidizing materials (chlorine) could be applied when dealing with soluble matter. Ozone could be also used as oxidizing agent. Ozone has higher redox potential than chlorine so the oxidizing state in reactions could be much easier reached [1].

The cyanide molecule is destroyed by oxidation. Chlorine is the oxidizing agent most frequently used. Oxidation must be conducted under alkaline conditions to avoid the generation of hydrogen cyanide gas. This process is often referred to as alkaline chlorination. This technology can be applied to a wide range of cyanide wastes: copper, zinc, and brass plating solutions; cyanide from cyanide salt heating baths; and passivating solutions. For extremely high cyanide concentrations (>1%), oxidation may not be desirable. Cyanide complexes of metals, particularly iron and to some extent nickel, cannot be decomposed easily by cyanide oxidation techniques.

Electrolytic oxidation of cyanide is carried out by anodic electrolysis at high temperatures. It has been more successful for wastes containing high concentrations of cyanide (50,000 to 100,000 mg l^{-1}), but is has also been successfully used for concentrations as low as 500 mg l^{-1}.

Chemical oxidation methods for organic compounds in wastewater have been extensively study. In general they apply only to dilute solutions and often are considered expensive in comparison to the biological methods. Some examples include wet air oxidation, hydrogen peroxide, permanganate, chlorine dioxide, chlorine, and ozone oxidation. Wet air oxidation, known as the Zimmerman process, based on the principle that most organic compounds can be oxidized by oxygen given sufficient temperature and pressure. This oxidation can be described as the aqueous phase oxidation of dissolved or suspended organic particles.

3.2.2 Solvent Recovery/Extraction

Waste treatment processes can use liquid-liquid or solvent extraction for removal of hazardous substances. Metal ions can be removed from wastewaters by liquid-liquid extraction processes involving ion exchange. Barium ions can be removed from wastewaters by treatment with a solution of dinonylnaphthalenesulfonic acid in heptane.

Solvent extraction is a step in a special process for converting sludges to very dry solids. With this process, water and oils can be removed from sludges by extraction with an aliphatic amine and separation of the solids from the liquid. The amine solution separates into organic and aqueous phases upon being heated, and the amine can be recovered and recycled [31].

3.3 Physico-Chemical Treatment Techniques

Chemical/physical treatment processes are those in which a chemical reaction is used to alter or destroy a hazardous waste component. Chemical treatment techniques can be applied to both organic and inorganic wastes, and may be formulated to address specific target compounds in a mixed waste. Typical chemical treatment processes include oxidation-reduction reactions such as ozonation, alkaline chlorination, electrolytic oxidation and chemical dechlorination. Physical treatment processes separate waste component by either applying physical force or changing the physical form of the waste. Various physical processes include adsorption, distillation, or filtration. Physical treatment is applicable to a wide variety of waste streams but further treatment is usually required.

Physical stabilization refers to improved engineering properties such as bearing capacity, trafficability, and permeability. Alteration of the physical character of the material to form a solid material reduces the accessibility of water to the contaminants within a cemented matrix and entraps or microencapsulates the contaminated solids within a dimensionally stable matrix. Since most of the contaminants in dredged material are tightly bound to the particulate fraction, physical stabilization is an important contaminant immobilization mechanism.

3.3.1
Adsorption on Activated Carbon

Adsorption is a mass transfer process in which gas vapors or chemicals in solution are held to a solid by intermolecular forces (for example, hydrogen bonding and Van der Waals' interactions). It is a surface phenomenon. Activated carbon, molecular sieves, silica gel, and activated aluminum are the most common adsorbents. The active sites become saturated at some point in time. When the organic material has commercial value, the bed is then regenerated by passing steam through it. The vapor-laden steam is condensed and the organic fraction is separated from the water. If the organic material has no commercial value, the carbon may be either incinerated or shipped to the manufacturer for regeneration. Carbon systems have been in commercial application for over 20 years.

The process of adsorption on carbon is applicable to wastewaters and polluted air. Treated waters may be suitable for reuse in industrial processes and can be discharged safely to the sewer system if removal efficiencies are high enough. Hazardous constituents of no commercial value that are removed may be disposed in burial sites after they have been stabilized. The high capacity of activated carbon for many compounds is attributable to the large surface area of the carbon (500–1500 $m^2\,g^{-1}$).

3.3.2
Distillation

The separation of more volatile materials from less volatile materials by a process of vaporization and condensation is called distillation. When a liquid mixture of two or more components is brought to the boiling point of the mixture, a vapor phase is created above the liquid phase. If the vapor pressure of the pure components is different (which is usually the case), then the constituent having the higher vapor pressure will be more concentrated in the vapor phase then the constituent having lower vapor pressure. If the vapor phase is cooled to yield a liquid, a partial separation of the constituents will result. The degree of separation depends on the relative differences in the vapor pressures. The larger the differences, the more efficient the separation. If the difference is not large enough, multiple cycles are required.

3.3.3
Exchange of Ions

Ion exchange columns are efficient for removing many different ions from aqueous solution. The ions exchanged may be either cationic or anionic.

Metals and ionized organic chemicals can be recovered by ion exchange. In ion exchange, the waste stream containing the ion to be removed is passed through a bed of resin. The resin is selected to remove either cations or anions. Ions are removed from the resin surface in exchange for ions in solution. Typically, either hydrogen or sodium is exchanged for cations (metal) in solution. When the bed becomes saturated with the exchanges ion, it is shut down and the resin is regenerated by passing a concentrated solution containing the original ion (hydrogen or sodium) back through the bed. A prefilter is required to remove suspended material that would hydraulically foul the column. It also removes organics and oils that would foul the resin. Ion exchange systems are suitable for chemical recovery applications where the rinse water feed has a relatively dilute concentration (<1000 mg l^{-1}). Ion exchange has been demonstrated commercially for recovery of plating chemicals from acid-copper, acid-zinc, nickel, tin, cobalt, and chromium plating baths.

The diameters of ion exchange columns may vary from 1 cm to 6 m. Resin bed depths range from 1 to 3 m. Bed height-to-diameter ratios range from 1.5:1 to 1:3. The column height generally does not exceed 4 m. Multiple columns in series are provided where the design height exceeds 4 m.

3.3.4 Reverse Osmosis

Osmosis is defined as the spontaneous transport of a solvent from a dilute solution to a concentrated solution across an ideal semipermeable membrane that impedes passage of the solute but allows the solvent to flow. Solvent flow can be reduced by exerting pressure on the solution side of the membrane. If the pressure is increased above the osmotic pressure on the solution side, the flow reverses. Pure solvent will then pass from the solution into the solvent. As applied to metal finishing wastewater, the solute is the metal and the solvent is pure water.

Many configurations of membrane are possible. The driving pressure is in the order of 2800 to 5500 kPa. No commercially available membrane polymer has demonstrated tolerance to all extreme chemical factors such as pH, strong oxidizing agents, and aromatic hydrocarbons. However, selected membranes have been demonstrated on nickel, copper, zinc, and chromium baths.

3.4 Thermal Treatment Techniques

Thermal treatment processes use energy to destroy or decontaminate waste. These technologies include low or high energy thermal processes. Several types of thermal processes include flame combustion, fluidized bed combustion, infrared incineration, pyrolysis and plasma heat systems.

3.4.1 Thermal Desorption/Destruction

The pyrolytic process also uses heat for the destruction of hazardous waste constituents. However, in contrast to incineration, the decomposition process includes an endothermic reaction which is performed at a lower temperature and in the absence of oxygen. General temperature range for the pyrolysis is between 425 and 750 °C. Pyrolysis is a two-stage process. At the first stage, heating of mixed waste at the lower temperature (425–750 °C) results in the separation of volatile from non-volatile fractions. At the second stage, volatile gasses are burning in the incinerator, with the remaining ash (solid residue). The two-stage process enables precise temperature control and demands smaller equipment.

Hazardous waste types, which are especially suitable for the pyrolytic treatment, include: containerized waste (i.e., barrels); and sludges and liquids that have high ash content, volatile inorganic compounds (NaCl, $FeCl_2$, Zn, and Pb), or high concentrations of chlorine, sulfur, and/or nitrogen.

3.4.2 Vitrification

Vitrification is a high temperature process of immobilizing, and chemically incorporating, radioactive and other hazardous wastes. The procedure uses high temperatures (typically between 1100 and 1600 °C). At these temperatures, waste material is transformed into an amorphous liquid. On cooling, the vitrification produces an amorphous, glass-like solid that permanently captures the waste. Extremely hazardous wastes and radioactive wastes can be immobilized by this method.

3.4.3 Plasma Heating

Various pesticides and PCBs can be degraded with high efficiency by the microwave plasma process when oxygen is used as the reactant gas. The plasma consists of ions, free electrons and neutral species. Highly energized electrons dissociate molecular oxygen into atomic oxygen, which can react rapidly with organic compounds. Also, organic molecules undergo dissociation into free radicals during bombardment with free electrons. These free radicals can react rapidly with oxygen.

3.4.4 Microwave Heating

There are two types of systems for the hazardous waste treatment with heating. Microwave furnaces with the operating area on the top are used for the treatment of small quantities of hazardous (including medical) waste. The second type is special devices for the microwave treatment. Waste is grinded, treated with vapor, and subjected to microwave radiation, all in one device. Waste is put in the hopper and could be fed chargely or continuously into the grinding chamber. The grinding process contributes to the volume reduction to about 80% [32].

3.4.5
Incineration

From the chemical point of view incineration represents exothermic oxidation process which converts organic compounds to carbon dioxide and water vapor with the liberation of heat. In an incinerator, chemicals are decomposed by oxidation at high temperatures (800 °C and greater). The waste, or at least its hazardous components, must be combustible in order to be destroyed. The primary products from combustion of organic wastes are carbon dioxide, water vapor, and inert ash. However, there are a multitude of other products that can be formed. Incineration is significant and useful way for the waste reduction. It is used for the treatment of hazardous and toxic waste, so as for municipal waste.

Waste which is charged to incinerator is not composed of organic compounds only. Inorganic materials include metals and glass which are often found in the waste. Metals and non-metals are submerged to oxidation as the result of incineration, e.g., metallic copper goes to copper oxide, sodium and potassium to hydroxide, fluorine to fluor-hydrogen, chlorine to chlor-hydrogen and carbon to carbon dioxide. Other constituents of waste are remained ash or flue gasses and require special handling, treatment, and disposal.

4
Mechanisms of Leaching Processes

The characterization of leaching behavior in water of solidified/stabilized waste is crucial in most reuse or disposal scenarios. Water plays a multiple role in physico-chemical phenomena in the solid, in pollutant transfer and pollutant dispersion in the environment.

4.1
Fundamentals of Leaching

The purpose of leaching tests is to obtain aqueous phase concentration of constituents which are released from solids when placed in a land disposal unit. Leachate concentration constitute the source term for transport and fate in the environment and, therefore, the associated risk. Leaching potential for the same chemical can be quite different depending on a number of factors such as characteristics of the leaching fluid, form of the chemical in the solids, and the disposal conditions.

4.2 Commonly Used Leaching Tests on Solidified/Stabilized Waste Products

There are a great number of leaching tests. The experimental conditions and interpretation of results are dependent on the application context and scientific objectives, as follows:

- *Application context of the tests (verification tests, compliance tests, and characterization tests).* Most of the compliance tests applied to solidification/stabilization waste are short term tests involving high liquid to solid ratios and which sometimes include crushing of the materials. Such tests are not appropriate to understanding leaching behavior or for the prediction of release from solidification/stabilization wastes.
- *Scientific objectives.* The three most common are: determination of the available release potential (this notion aims to evaluate the total elementary content of pollutant available for leaching), study of solubilities at equilibrium (these tests which are conducted on crushed waste are carried out either at controlled pH or by addition of given quantities of acid when measuring the acid neutralization capacity, ANC), and assessment of the release dynamics (these tests are carried out on monolithic samples over long periods from one to three months or more) [33–36].

4.3 Leaching Studies

4.3.1 Column

Column test (described in Dutch standard NEN 7343) is intended to simulate the percolation of (acidic) rainwater through granular materials in a road construction or landfill. The material is placed in a column with a diameter of 5 cm and height of 20 cm, and acidified demineralized water is then passed upwards through it. The eluate is collected in seven fractions at a liquid/solid ratio of between 0.1 and 10 l kg^{-1}. The leaching is calculated and expressed in mg kg^{-1} as a function of time, based on the analysis results.

4.3.2 Batch

This test, which lasts 64 days and is described in Dutch standard NEN 7345, is intended to determine leaching from products or stabilized waste materials as a function of time. The product is immersed in acidified demineralized water. This eluate is refreshed at seven specified times. Analysis of the eluates enables the leaching per square meter of product surface area and the effective diffusion coefficient to be calculated.

4.3.3 Cascade

Cascade test (described in Dutch standard NEN 7349) provides information on long-term leaching from a granular material. In this test the material is brought into contact with acidified demineralized water five times in succession for 23 h. The cumulative L/S (liquid/solid) ratio is 100 l kg^{-1}. This test also yields an emission, expressed in mg kg^{-1}.

4.3.4 TCLP

Toxicity Characteristic Leaching Procedure, Method 1311 (TCLP) is a standard batch leaching test developed by US EPA for classification of waste. The test which replaced the EP Toxicity test in US regulation is designed to determine the mobility of inorganic and organic contaminants in liquid, solid and multiphase waste. Leaching is carried out at L/S=20 l/kg with the contact time of 24 h. The most significant difference is that the TCLP prescribes the choice between two leachants, depending on the alkalinity of waste material.

4.3.5 GANC

A method is developed to measure the neutralizing capacity of various generic binders. The method, termed the generalized acid neutralization capacity (GANC) test, is based on modifications to Environment Canada Method No.7, Acid Neutralization Capacity, combined with US EPA Method No. 1311, Toxicity Characteristic Leaching Procedure. The GANC test is a single batch procedure that utilizes leachants normally ranging in acidity from 0 to 6 equivalents of acid per kilogram of solids. After 48 h of contact, the pH of the leachate is measured [37].

4.4 Parameters of Leaching Processes

The different parameters that must be considered in leaching tests are: leaching time; liquid/solid ratio; sample granulometry (i.e., monolithic solid or crushed powder); leachant (composition, pH, temperature, etc); and renewal conditions of the leachant. For most inorganic compounds, the controlling factors for leaching are: pH; redox conditions; liquid/solid ratio for extraction; solubility; and solid phase compound. On the other hand, factors controlling leaching behavior of organic compounds are: solubility; partitioning; presence of organic carbon; liquid/solid ratio for extraction; and non-aqueous phase liquid. Typical organics treated by solidification/stabilization include trichloroethene, benzene, methylene chloride, pentachlorophenol, etc.

4.5 Physico-Chemical Mechanisms of Mass Transfer

The pollutant flux due to leaching depends on the physico-chemical mechanisms which govern mass transfer when solidified/stabilized waste comes into contact with water in the scenario conditions. The use of hydraulic binders has several effects on the leaching behavior of pollutants (mainly metallic species) including [22]:

- Alkalinity with formation of slightly soluble hydroxides (except for amphoteric metals)
- Formation of slightly soluble silicates
- Insertion of certain metals in certain phases of the hydrates, in particular by substitution of calcium or sulfate ions in the ettringite
- Absorption at the CSH surface

The release of soluble species contained in a porous medium in contact with water is the result of complex and coupled phenomena. This includes water transfer in the porous medium up to saturation; dissolution of the species in the pore water according to the local chemical context; and transport of species in solution due to the effect of concentration gradients.

Potential change of species solubility in the pore water if the context of the pore water has undergone certain modifications due to the pH profile for example, following the release of pH controlling species.

4.6 Data Interpretation of Leaching Tests

Whereas certain models only take into account Fick's diffusion law, by proposing the "analytical" solution or the numerical one, other models take into account a coupling effect like diffusion/chemical reaction, absorption and or/dissolution, or transfer at the interface. Numerous research programs have been conducted on the interpretation of leaching tests to develop tools, for the evaluation of long term release of soluble pollutants contained in solidified/stabilized waste. From the obtained experimental results, two cases may occur: (a) relatively soluble elements whose solubility does not change according to the variable physico-chemical leaching context; and (b) elements whose solubility depends on the variation of the physico-chemical context, pH in particular.

4.7 Environmental Impact Assessment on Surface and Ground Waters

The prevention of contamination of surface and ground waters by leaching from hazardous waste is a key of best environmental practice. The dumping of untreated hazardous waste represents a very real threat to a quality of surface and ground waters. This limited and valuable resource must be protected from contamination and hence cheap, simple technologies which eliminate leaching from hazardous wastes need to be developed and promoted [20].

5
A Case Study: Solidification/Stabilization of Lead and Cadmium Ions in the Cement Matrix

5.1
Overview

This case study represents an unusual case in a galvanic treatment plant in Serbia. The goal was to find out a way to stabilize sludge containing mainly lead and cadmium. Portland cement and Portland cement with 30% of fly ash were added as binders for immobilization of Pb^{2+} and Cd^{2+} ions. Samples were prepared by mixing cement with sludge from the galvanization containing: 10,000 mg l^{-1} Pb^{2+}, 30,000 mg l^{-1} Pb^{2+}, and 50,000 mg l^{-1} Cd^{2+}. Those samples were immersed into an aggressive acid solution (pH=4) and into deionized water (used as reference). The temperature of the solutions was 20 °C and 50 °C, respectively. Flexural strength of samples was measured. After 1, 3, 7, 14, 28, 35, 42, 49, and 56 days the concentration of toxic metal ions and Ca^{2+} ions in leachate solutions was determined by atomic absorption spectrometry.

5.2
Experimental Work

Portland cement and Portland cement with 30% fly ash were used as matrix. Chemical composition, physico-chemical and mechanical properties of cements were determined. The samples 1×1×6 cm were prepared by mixing cement with wastewater containing 10,000 mg l^{-1} Pb^{2+}, 30,000 mg l^{-1} Pb^{2+}, and 50,000 mg l^{-1} Cd^{2+}. The next samples were prepared:

- Cement paste 10: Portland cement (OPC)+distilled water
- Cement paste 11: OPC+10,000 mg l^{-1} Pb^{2+}
- Cement paste 13: OPC+30,000 mg l^{-1} Pb^{2+}
- Cement paste 15: OPC+50,000 mg l^{-1} Cd^{2+}
- Cement paste 20: Portland fly ash cement (PPC)+distilled water
- Cement paste 21: PPC+10,000 mg l^{-1} Pb^{2+}
- Cement paste 23: PPC+30,000 mg l^{-1} Pb^{2+}
- Cement paste 25: PPC+50,000 mg l^{-1} Cd^{2+}

Those samples were immersed in the aggressive acid solution (pH=4) and deionized water as reference. The temperatures of acid solution and deionized water were 20 °C and 50 °C. Flexural strength of samples was determined at the beginning and after 56 days in aggressive solution. After 1, 3, 7, 14, 28, 35, 42, 49, and 56 days the concentration of toxic metal ions and Ca^{2+} ions in leachate solutions was determined by atomic absorption spectrometry. Changes into microstructure were determined by scanning electron microscopy and energy dispersive analysis.

5.3 Results and Discussion

Portland cement clinker potential phase composition is presented in Table 4. It could be seen that the C_3A content in the clinker was 9.46% which is important for the cement hydration rate and cement sulfate resistance. Common Portland cement is not resistant to the sulfate influence because of the significant C_3A content, whose hydrates react with sulfate ions resulting in expansive compounds. Portland cement with the higher resistance to sulfates must have low C_3A content. Moderate to high content of mineral alite – C_3S (54.72%) is usual for the Serbian cement plants and enables the addition of higher quantities of mineral admixtures without influencing the quality of final cement.

Chemical composition of the mineral admixture – coal burning fly ash is presented in Table 5. Because of the lower content of SiO_2 and high content of CaO this fly ash was not so suitable for the cement production. It also has the increased loss on ignition, yet complying with the Yugoslav standard JUS B.C1.018. It also could be seen that the MgO, alkalies, and SO_3 content were in the limits of the mentioned Yugoslav standard.

Pozzolanic activity of the used fly ash sample (Table 6), determined by the test with the hydrated lime according to the mentioned standard, is for the P5 pozzolana class which means that it is convenient for the cement production as the mineral admixture [38–41].

Both cements (Table 7) compile with the requirements of the Yugoslav standard JUS B.C1.011, concerning chemical composition. Portland fly ash cement PPC has the expected higher insoluble residue and the loss on ignition regarding Portland cement OPC because of the fly ash addition. Other constituents were not significantly changed with the fly ash addition.

Fly ash addition was obviously (Table 8) increasing the demand for the water for the standard consistence and prolonging the setting time, but has no influence on the other cement characteristics. All characteristics were in the compliance with the Yugoslav standard JUS B.C1.011.

Table 4 Potential phase composition of Portland cement clinker

Potential phase composition, mass %	Portland cement clinker
C_3S	54.72
C_2S	18.47
C_3A	9.46
C_4AF	8.49

Table 5 Chemical composition of fly ash

Chemical composition, mass %	Fly ash
LOI	8.81
SiO_2	36.68
Al_2O_3	13.38
Fe_2O_3	4.78
CaO	29.44
MgO	2.82
SO_3	2.70
Na_2O	0.40
K_2O	0.97

Table 6 Pozzolanic activity of fly ash

	Flexural strength, MPa	Compressive strength, MPa
Pozzolanic activity of fly ash	1.3	6.9

Table 7 Chemical composition of Portland cement and Portland fly ash cement

Chemical composition, mass %	Cement	
	OPC	PPC
SiO_2	20.20	19.98
Al_2O_3	5.10	5.61
Fe_2O_3	2.39	2.39
CaO	62.80	55.23
Insoluble residue	0.65	6.85
Moisture at 105 °C	0.90	1.40
LOI	1.79	2.91
Free CaO	0.93	1.24
CaO in $CaSO_4$	3.48	1.67
SO_3 in $CaSO_4$	2.55	2.39
MgO	3.02	3.62
Alkalies as Na_2O	0.36	0.32
K_2O	0.74	0.70
Cl^-	0.00	0.00

Table 8 Physico-chemical properties of Portland and Portland fly ash cement

Physico-chemical properties	Cement	
	OPC	PPC
Sieve residue at 0.09 mm, % mass	1.4	1.6
Specific surface, cm^2g^{-1}	4000	4500
Setting		
Standard consistence, mass %	25.50	28.25
Initial time, min	55	125
Final time, min	145	205
Soundness		
Le Chatelier test, mm	1.0	1.0

Values for cement flexural and compressive strengths after 3, 7, and 28 days are presented in Table 9. Portland fly ash cement PPC had lower strengths in all ages regarding corresponding Portland cement OPC which is normal consequence of the fly ash addition. Namely, according to the literature the addition of fly ash is slowing down Portland cement hydration in early ages, but later (at the age of 90 days and later) strengths of such cements are equalizing or become much higher than strengths of the corresponding Portland cement. In our case both cements complies requirements of the mentioned standard for the class 45. Flexural strengths of samples exposed to aggressive media were measured after 56 days. It could be concluded that heavy metal ions presence had no influence on strengths in such aggressive media [42].

Lead leaching concentration from the samples prepared with 10,000 mg l^{-1} Pb^{2+}, Figs. 1 and 2, was not above 0.1 mg l^{-1}. Higher leaching was in the acid media, but also below 0.1 mg l^{-1}, with samples with Portland cement regarding

Table 9 Strengths of Portland cement and Portland fly ash cement

Strengths, MPa	Cement	
	OPC	PPC
Flexural strength		
3 days	6.9	5.9
7 days	7.6	7.8
28 days	9.2	8.3
Compressive strength		
3 days	36.6	25.5
7 days	43.1	31.0
28 days	52.0	40.7

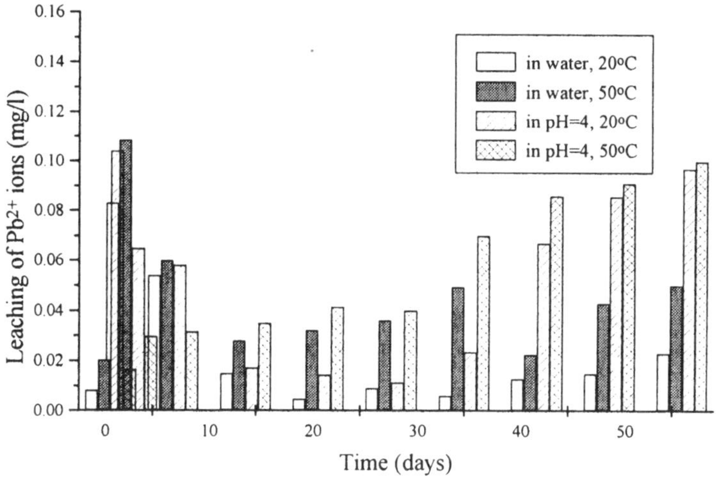

Fig. 1 Leaching of Pb^{2+} ions from the Portland cement sample 11

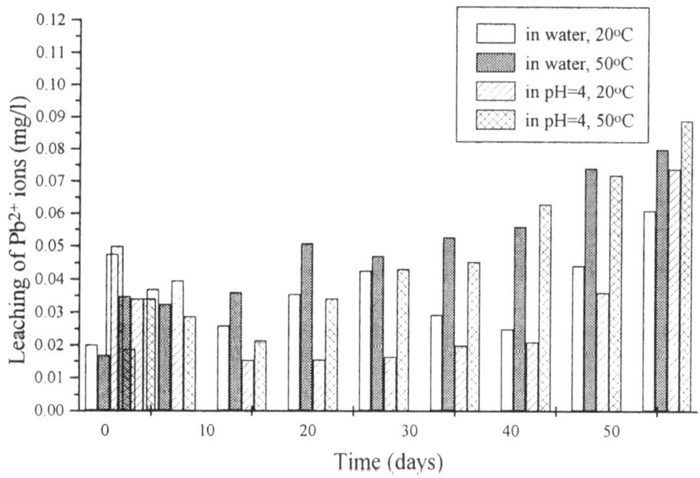

Fig. 2 Leaching of Pb^{2+} ions from the Portland fly ash cement sample 21

Portland fly ash cement samples, which was expected because it is known that the fly ash addition increases the resistance to acid media. It could be said that the lead in the concentration of 10,000 mg l^{-1} was successfully stabilized in the Portland cement and Portland fly ash cement matrix [43].

Figures 3 and 4 show lead leaching results from the samples prepared with 30,000 mg l^{-1} of Pb^{2+}. It could be seen that the leaching was over 0.1 mg l^{-1}, which is maximum allowed concentration for water streams, which means that the lead with this concentration couldn't be immobilized in such a way that

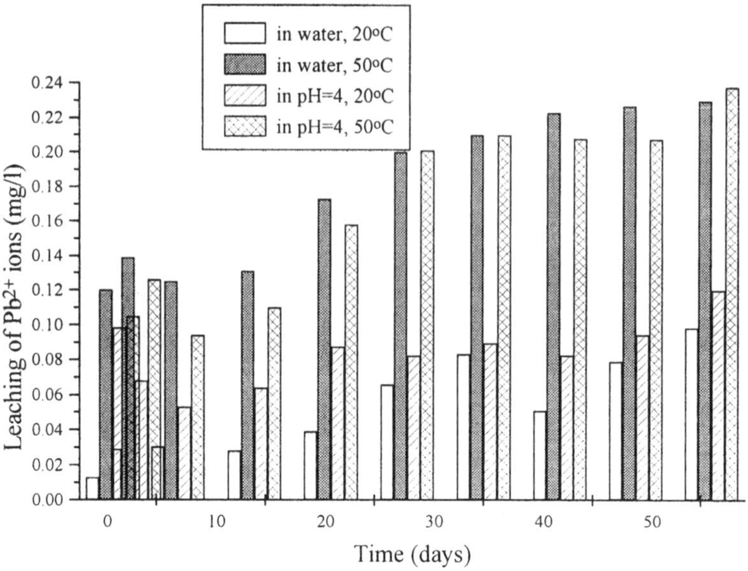

Fig. 3 Leaching of Pb^{2+} ions from the Portland cement sample 13

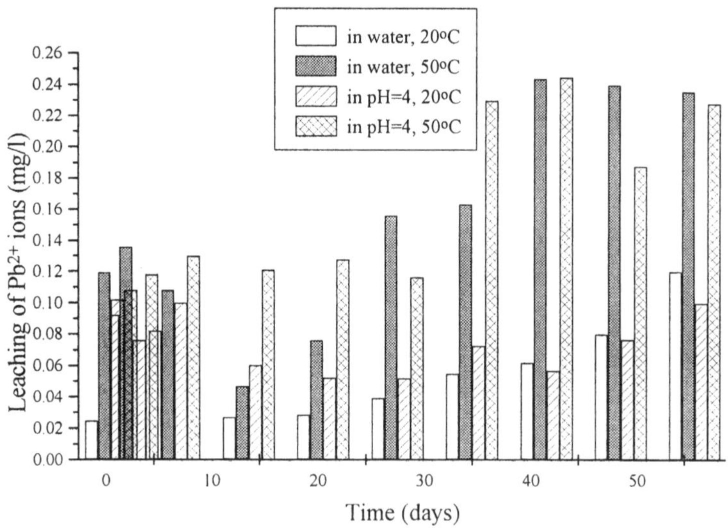

Fig. 4 Leaching of Pb^{2+} ions from the Portland fly ash cement sample 23

there is no danger of pollution, if such material is to be used in civil engineering (roads, embankments, foundations, etc.)

Figures 5 and 6 show cadmium leaching results from the prepared samples. There was the deviation of decreasing of cadmium concentration in the media during time.

This means that in the static conditions cadmium diffusion to solution mechanism was not prevailing, but the adsorption/diffusion/solution mechanism was present [22]. This means that the cadmium leached produced insol-

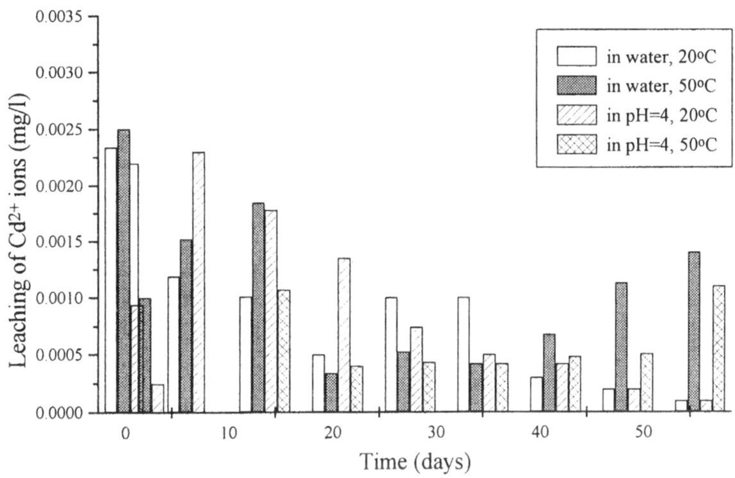

Fig. 5 Leaching of Cd^{2+} ions from the Portland cement sample 15

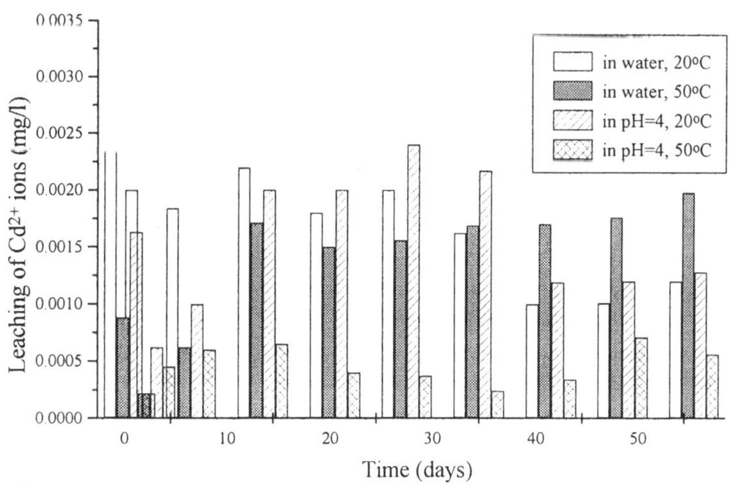

Fig. 6 Leaching of Cd^{2+} ions from the Portland fly ash cement sample 25

uble compounds in the solution which adsorbed at the cement paste. In any case, cadmium leaching was below 0.005 mg l^{-1}, which shown that the cadmium was successfully immobilized in the cement matrix [44].

Leaching of calcium from cement is, generally, somewhat increased if we compare water- to acid-exposed samples. Also, leaching of calcium is increased at 50 °C (compared to 20 °C). Leaching of calcium does not depend from lead ions immobilization (Figs. 7–10).

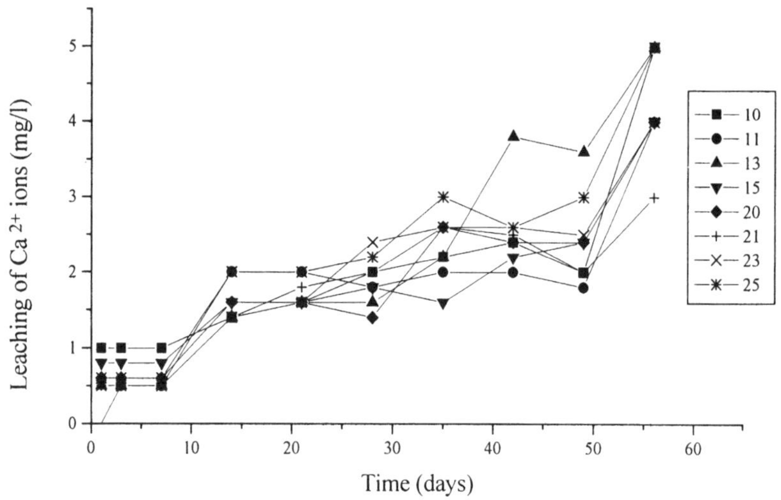

Fig. 7 Leaching of Ca^{2+} ions in distilled water at 20 °C

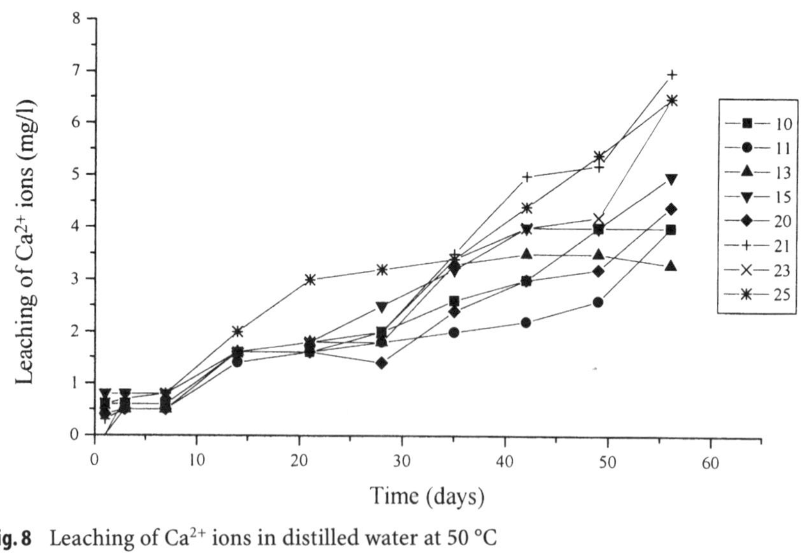

Fig. 8 Leaching of Ca^{2+} ions in distilled water at 50 °C

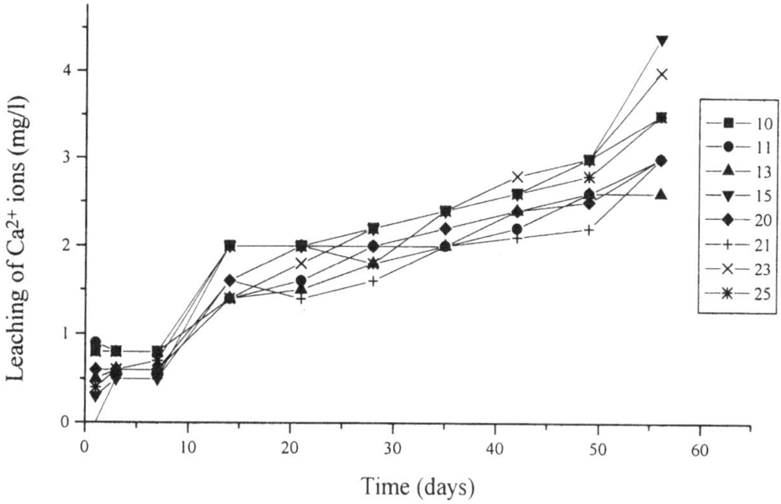

Fig. 9 Leaching of Ca^{2+} ions in pH=4, at 20 °C

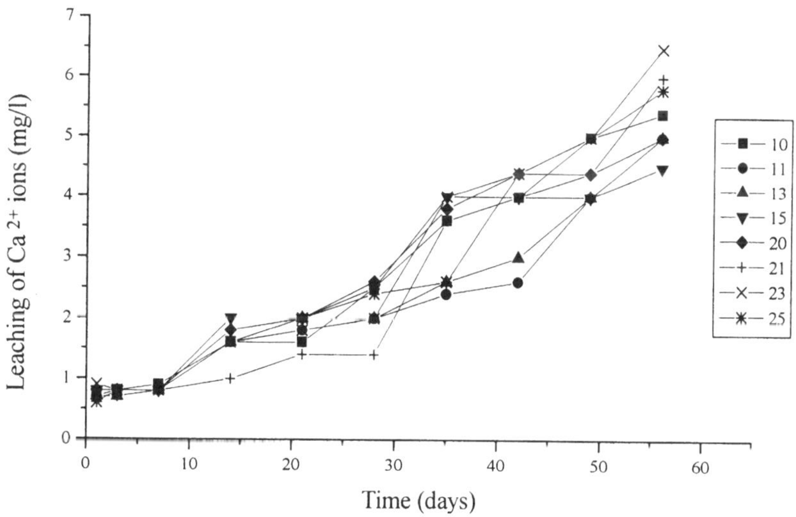

Fig. 10 Leaching of Ca^{2+} ions in pH=4, at 50 °C

5.4 Summary of Findings

The following is a summary of the most important findings in the present case study:

- Successful stabilization/solidification of cadmium in the concentration 50,000 mg l^{-1} of Cd^{2+}, and lead in the concentration 10,000 mg l^{-1} of Pb^{2+}, in the cement matrix, was obtained. Low concentrations during leaching pointed out that the heavy metal containing waste could be successfully stabilized and immobilized on this way and landfilled, or safely utilized as the building material.
- Lead immobilization with the concentration 30,000 mg l^{-1} of Pb^{2+} was not satisfying regarding maximum allowed concentration of lead for waterstreams, which means that it couldn't be used in civil engineering (roads, embankments, foundations, etc.), but could be safely landfilled at the landfill for the hazardous waste.
- Both Portland cement and Portland cement with the addition of 30% of fly ash enabled good immobilization, but Portland cement showed somewhat better results.
- The presence of lead and cadmium in the cement matrix with those concentrations did not lead to decreasing of samples strengths. Lead showed better improvement in stabilization and then cadmium. Lead dissolution at low pH values corresponded closely with the loss of aluminum, suggesting an ettringite or ferrite stabilization mechanism. The leachability of cadmium is continuous as the pH decreases from 9.5 to lower values during the batch leaching steps, and confirms a simple insoluble hydroxide stabilization mechanism and pH-controlled dissolution.

6 Conclusions and Recommendations

The issue of proper management of hazardous wastes is one which suffers from much misinformation and confusion. The present chapter reviewed additives and techniques that can be applied to specific solidification problems and immobilization of specific hazardous constituents (e.g., lead and cadmium), including a list of generic additives that can be used to control wastes pH; to reduce, oxidize, and co-precipitate constituents; and to accelerate/retard set.

Solidification/stabilization processes play a more important role since, in the near future, only inert or stabilized wastes should be landfilled. Solidification/stabilization process means binding the hazardous material in the hydraulic binders for safe landfilling or use in civil engineering purposes. Various types of cement, inorganic binders and pozzolanas (e.g., coal burning fly ash, lime, blast-furnace slag and similar materials) are mostly used as the stabilizing matrix. Those stabilization techniques are used for the immobilization of inorganic or organic waste. The end product of the treatment, usually after sufficient curing, is solid monolithic material which, depending on characteristics of leaching, can be usefully applied or disposed of in a safe way.

References

1. Davis M, Cornwell D (1986) Introduction to environmental engineering, 3rd edn. McGraw-Hill
2. Barlishen KD, Baetz BW (1996) Development of a decision support system for municipal solid waste management systems planning. Waste Manage Res 14:71–86
3. Watts RJ (1998) Hazardous wastes: sources, pathways, receptors. 0-471-00238-0
4. Spence RD (ed) (1993) Chemistry and microstructure of solidified waste forms. Lewis Publishers, Boca Raton, FL
5. Wilk CM (1998) Portland cement-based solidification/stabilization treatment of waste. Proceedings of the 4th International Symposium and Exhibition on Environmental Contamination in CEE, 4 pp
6. Cocke DL (1994) Recent advances in stabilization and solidification. Proceedings of the Spring National Meeting of the American Chemical Society. San Diego, CA, pp 537–541
7. Dawson GW, Mercer BW (1986) Hazardous waste management. 0-471-82268-X
8. Sell NJ (1992) Industrial pollution control: issues and techniques, 2nd edn. 0-471-28419-X
9. Woodside G (1999) Hazardous materials and hazardous waste management, 2nd edn. Wiley, 0-471-17449-1
10. Zheng C, Bennett GD (1995) Applied contaminant transport modelling: theory and practice. Wiley. 0-471-28536-6
11. Turnberg WL (1996) Biohazardous waste: risk assessment, policy, and management. Wiley. 0-471-59421-0
12. Albino V and others (1996) Potential application of Ettringite generating systems for hazardous waste stabilization. J Hazard Mater 51:241–252
13. Cheng KY, Bishop P (1992) Metals distribution in solidified/stabilized waste forms after leaching. Hazard Waste Hazard Mater 9:163–171
14. Wilk CM (1995) Portland cement gives concrete support to solidification/stabilization, Env Solutions
15. Mc Carthy GJ (1992) Synthesis, crystal chemistry and stability of Ettringite, a material with potential applications in hazardous waste immobilization. Proc Mater Res Socy Sympos 245:129–140
16. Owens JW, Stewart S (1996) Cement binders for organic wastes. Mag Concr Res 48:37–44
17. Roy A, Cartledge FK (1997) Fundamental aspects of cement solidification and stabilization. J Hazard Mater 52:151–154
18. Macphee DE, Glasser FP (1993) Immobilization science of cement systems. MRS Bull 3:66–71
19. Roy A (1992) Solidification/stabilization of hazardous waste: evidence of physical encapsulation. Environ Sci Technol 26:1349–1353
20. Mollah YMA (1995) The interfacial chemistry of solidification/stabilization of metals. Waste Manage 15:137–148
21. Conner JR, Hoeffner SL (1998) The history of stabilization/solidification technology. Crit Rev Environ Sci Technol 28:325–396

22. Bonen D, Sarkar SL (1995) The effects of simulated environmental attack on immobilization of heavy metals doped in cement-based materials. J Hazard Mater 40:321–335
23. Weng CH, Huang CP (1994) Treatment of metal industrial wastewater by flyash and cement fixation. J Environ Eng 120:1470–1488
24. Ilic M, Jaksic B (2000) Hazardous waste management. Institute of Urbanism. Republic of Srpska, ISBN 86-7440-004-3
25. Meyers RA, Kender Dittrick D (eds) (1999) Encyclopedia of environmental pollution and cleanup. Wiley
26. Tucker SP, CarsonGA (1985) Deactivation of hazardous chemical wastes. Environ Sci Technol19:215–220
27. EPA 530-R-97-057 (1997) Introduction to hazardous waste incinerators. EPA
28. Murck BW, Skinner BJ, Porter SC (1995) Environmental geology. Wiley
29. Nocun-Wczelik W, Malolepszy J (1997) Study on the immobilization of heavy metals in cement paste – C-S-H leaching behaviour Proceedings of the 10th ICCC, Göteborg, Sweden, vol. 4, 4iv043, 8 pp
30. Albino V, Cioffi R, Santoro L, Valenti GL (1996) Stabilization of residue containing heavy metals by means of matrices generating calcium trisulphoaluminate and silicate hydrates. Waste Manage Res 14:29–42
31. Ludwig C, Ziegler F, Johnson CA (1997) Heavy metal binding mechanisms in cement based waste materials. In: Goumans JJJM, Senden Y, Van der Sloot HA (eds) Waste materials in construction – putting theory into practice. Studies in environmental science 71, Elsevier Science, pp 355–364
32. Gougar MLD, Scheetz BE, Roy DM (1996) Ettringite and C-S-H Portland cement phases for waste ion immobilization: a review. Waste Manage 16:295–303
33. Hasan S (1996) Geology and hazardous waste management. Prentice Hall
34. Isenburg J, Moore M (1992) Generalized acid neutralization capacity test. In: Gilliam TM, Wiles CC (eds) Stabilization and solidification of hazardous, radioactive, and mixed wastes, vol 2. ASTM STP 1123, American Society for Testing and Materials, Philadelphia, pp 361–377
35. Stegeman JA, Caldwell RJ, Shi C (1996) Laboratory, regulatory, and field leaching of solidified waste. In: Proceedings of the International Conference on Incineration and Thermal Treatment Technologies, Savannah, GA, pp 75–80
36. Lee CH (1994) A long-term leachability study of solidified wastes by the multiple toxicity characteristic leaching procedure. J Hazard Mater 38:65–74
37. Trussell S, Spence RD (1994) A review of solidification/stabilization interferences. Waste Manage 14:507–519
38. Stegemann JA, Cote PL (1996) A proposed protocol for evaluation of solidified wastes. Sci Total Environ 178:103–110
39. Albino V (1996) Evaluation of solid waste stabilization processes by means of leaching tests. Environ Technol 17:309–315
40. Cartledge FK, Yang SL (1990) Immobilization mechanisms in solidification/stabilization of cadmium and lead salts using Portland cement fixing agents. Environ Sci Technol 24:867–873
41. Wang SY, Vipulanandan C (1996) Leachability of lead from solidified cement-fly ash binders. Cem Concr Res 26:895–905
42. Poon CS, Lio KW (1997) The limitation of the toxicity characteristic leaching procedure for evaluating. cement-based stabilized/solidified waste forms. Waste Manage 17: 15–23
43. Van der Sloot HA, Heasman L, Quevauviller P (1997) Harmonization of leaching/extraction tests. Studies in environmental science 70, Elsevier Science

44. Stegeman JA and others (1995) Quality analysis of field solidified waste. Proceedings of the 5th Annual Symposium on Groundwater and Soil Remediation, Toronto, Canada, pp 553–555
45. Hills C (1993) Ordinary Portland cement based solidification of toxic wastes: the role of OPC reviewed. Cem Concr Res 23:196–212
46. Risch A, Pollmann H, Ecker M (1997) Application of Portland cement and high alumina cement for immobilization/solidification of a waste model composition. Proceedings of the 10th ICCC, Göteborg, Sweden, vol. 4, 4iv044, 8 pp
47. Ilic M, Miletic S, Milic D, Brceski I (2000) Solidification of lead ions in Portland cement matrix. In: Woolley GR, Goumans JJJM, Wainwright PJ (eds) Waste materials in construction – science and engineering of recycling for environmental protection. Waste management series, Elsevier Science, pp 87–97
48. Ilic M, Miletic S, Brceski I, Milic D (1999) Solidification/stabilization of toxic metal ions in Portland cement and Portland fly ash cement. Proceedings of Sardinia 99, 7th International Waste Management and Landfill Symposium, S.Margarita di Pula (Cagliari), Italy, pp 465–472

Handb Environ Chem Vol. 5, Part F, Vol. 3 (2005): 191–229
DOI 10.1007/b98267

Waste Minimization and Molecular Nanotechnology: Toward Total Environmental Sustainability

Tarek A. Kassim (✉)

Department of Civil and Environmental Engineering, Seattle University, 901 12th Avenue, P.O. Box 222000, Seattle, WA 98122–1090, USA
Kassimt@seattleu.edu

1 **Introduction** . . . 192

2 **Waste Minimization** . . . 194
2.1 "End-of-Pipe" Pollution Control Approaches . . . 195
2.2 Waste Management Hierarchy . . . 197
2.2.1 Waste Minimization . . . 197
2.2.2 Source Reduction . . . 198
2.2.3 Recycling . . . 199
2.3 Incentives and Strategies . . . 199
2.4 Recycling Techniques . . . 200
2.5 Barriers . . . 202
2.6 Case Study . . . 203
2.6.1 Source Reduction . . . 203
2.6.2 Recycling . . . 208

3 **Molecular Nanotechnology** . . . 209
3.1 Theory . . . 209
3.2 Environmental Benefits . . . 210

4 **Total Environmental Sustainability** . . . 211
4.1 Current Human Impact . . . 212
4.2 Environmental Protection and Remediation . . . 214
4.2.1 Pollution Control at Source . . . 215
4.2.2 Remediation of Existing Problems . . . 220

5 **Conclusions** . . . 224

References . . . 225

Abstract The issue of clean production has challenged many industries to initiate new approaches to tackle pollution problems. The traditional "end-of-pipe" treatment approach is no longer viewed as an adequate, stand-alone, pollution problem-solver. Increasing public awareness of the impact of industrial pollution, more stringent discharge standards, and escalating waste treatment and disposal costs have placed enormous pressure on industries to shift their paradigm of pollution prevention from the "end-of-pipe" treatment to waste minimization or even total elimination at the point of generation (source reduction). By reducing the energy and materials required to provide goods and services, an emerging new technology such as molecular nanotechnology (MNT) has the potential to provide more appealing products while improving environmental performance and total sustainability.

Therefore, the goals of this chapter are to: (a) provide an overview of waste minimization and its relationship to environmental sustainability; (b) portray the causes of sustainability problems and diagnose the defects of current industrial manufacturing processes in light of MNT, and; (c) analyze and extrapolate the prospect of additional capabilities that humanity may gain from the development of MNT that have the potential to ascertain total environmental sustainability.

Keywords Waste minimization · Sustainability · Molecular nanotechnology · Top-down bulk technology · Technology-to-date · Solid waste materials

Abbreviations

API	American Petroleum Institute
BTU	British Thermal Unit
CFCs	Chlorofluorocarbons
CO_2	Carbon dioxide
MNT	Molecular nanotechnology
mRNA	Messenger RNA
MSDS	Material safety data sheet
NOx	Nitrogen oxides
ODP	Ozone depletion potential
ODS	Ozone depleting substances
PCBs	Polychlorinated biphenyls
R&D	Research and development
RCRA	Resource Conservation and Recovery Act
SO_2	Sulfur dioxide
SWMs	Solid waste materials
TDBT	Top-down bulk technology
THMs	Trihalomethanes
TTD	Technology-to-date

1 Introduction

Reduction of hazardous wastes through prevention (i.e., source reduction) and recycling is an idea whose time has come. The general public, scientific and engineering communities, and government agencies are seeking solutions to the various problems presented by hazardous wastes [1, 2]. In the past, many governments have responded to evidence of damage to human health and the environment by regulating the treatment and disposal of hazardous wastes, and attempting to mitigate damage already done [3–14]. In recent years, however, widespread concern over risks to public health and damage to the environment has pressured many national and international organizations and governments to adopt ways of limiting the generation of hazardous wastes at their sources [15–23]. For example, a national USA policy of reducing or eliminating the generation of hazardous wastes was promulgated in the 1984 amendments to RCRA (Resource Conservation and Recovery Act). These amendments recommended that hazardous waste reduction at the source be considered first – before recycling, treatment and disposal options [18, 21, 22].

Minimizing the generation of hazardous waste is a promising alternative for industry because the costs of treating and disposing of hazardous wastes continue to rise. As a result, many industrial businesses that have adopted waste minimization programs have reduced operational costs, improved worker safety and enhanced their public image [24–27]. However, one of the factors identified as a barrier to wider participation of industry in waste minimization programs has been a lack of information about practical alternatives to treatment and disposal [5, 26]. In general, the waste minimization approach differs from treatment and disposal [16, 17, 23, 26] in that: (a) it focuses first on preventing the generation of waste (source reduction) and second on recycling any waste generated; (b) its successful application relies on an assessment of the overall and individual steps of manufacturing, processing or handling operations, and; (c) it succeeds best with strong endorsement from management, and active support and participation throughout the industrial facility.

In general, the relationship between technologies and the environment is very complex. With current state-of-art technologies, humans are threatened by environmental pollution and hazardous wastes [1, 2]. Hence, environmental sustainability based on technology-to-date (TTD) is pessimistic [1, 26, 28–30]. In contrast, emerging molecular nanotechnology (MNT) in all industrial fronts (for instance, nano-electronics, nano-biotechnology, nano-material, nano-energy) offers radical tools that may enable human society to gain the upper hand in the struggle toward total environmental sustainability and economic growth for the first time [31–34]. MNT will have extra benefits for human civilization beyond the remediation of environmental liabilities accumulated since the Industrial Revolution of the eighteenth century, providing the means to produce unlimited materials and energy with ultra green processes (pollution free [35]).

Accordingly, one of the main goals of the present chapter is to provide an overview of waste minimization and its relationship to environmental sustainability. This extensive evaluation will: (a) underline the inadequacy of so called "end-of-pipe" pollution control approaches by reviewing several examples of global and local environmental pollution problems threatening human health and the integrity of the environment; (b) discuss the waste management hierarchy which includes source reduction, onsite and offsite recycling, and treatment/disposal; (c) report the incentives gained from adopting waste minimization practices; (d) discuss the barriers discouraging the adoption of source reduction and recycling by industry, and; (e) present a case study from the drilling industry, where some businesses have successfully applied waste minimization practices. In addition, this chapter portrays the causes of sustainability problems and diagnoses the defects of current industrial manufacturing processes in light of molecular nanotechnology. It also analyzes and extrapolates the prospect of additional benefits that humanity may gain from the development of nanotechnology, that has the potential to create total environmental sustainability.

2
Waste Minimization

In general, the term "hazardous waste" is defined as any discarded material that may pose a substantial threat or potential hazard to human health or the environment when managed improperly [16, 17, 23]. According to RCRA, the term "solid waste" means any garbage, refuse, sludge, from a waste treatment plant, water supply treatment plant or air pollution control facility and other discarded material, including solid, liquid, semi-solid, or contained gaseous material resulting from industrial, commercial, mining and agriculture operations and from community activities [18, 20, 22].

The United States EPA, in implementing RCRA, has further clarified and broadened the definition of "solid waste" in a series of definitions and sub-definitions [21, 26], as follows:

- *Solid Waste*: Any discarded material not excluded by regulation (article # 40CFR 261.2(a))
- *Discarded Material*: Material abandoned, recycled or inherently waste-like (article # 40CFR 261.2(c))
- *Abandoned*: Disposed of, burned, or incinerated, or accumulated, stored or treated prior to or in lieu of abandonment (article # 40CFR).

On the other hand, RCRA defines "hazardous waste" as a solid waste or combination of solid wastes [1, 18, 20, 22], which because of its quantity, concentration, or physical, chemical or infectious characteristics may: (a) cause, or significantly contribute to an increase in mortality or an increase in serious

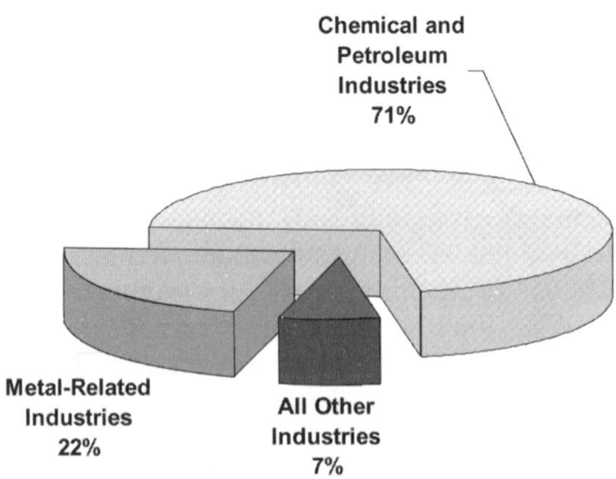

Fig. 1 Major sources of industrial hazardous wastes

irreversible, or incapacitating reversible, illness; or (b) pose a substantial present or potential hazard to human health or the environment when improperly treated, stored, transported, or disposed of, or otherwise managed.

While about 850,000 different industrial plants across the U.S. contribute to the total amount of hazardous waste generated, most of the industrial hazardous waste is mainly produced by the chemical and petroleum (71%), and metal-related (22%) industries (Fig. 1). Hazardous wastes are generated by both large and small businesses. Small businesses, although not individually generating large volumes of wastes, often generate highly toxic and more concentrated waste [36, 37].

The next paragraphs discuss the inadequacy of "end-of-pipe" pollution control approaches, waste management hierarchy, incentives gained from adopting waste minimization practices, barriers to effective waste minimization, and applications of waste minimization programs in drilling operations. The following is a summary.

2.1 "End-of-Pipe" Pollution Control Approaches

The management of hazardous wastes exclusively through "end-of-pipe" pollution control approaches has proved inadequate for safe-guarding human health and preventing environmental degradation [16, 17, 24, 26, 36, 37]. In spite of numerous laws and regulations, and extensive government and industry expenditures for treatment and disposal of pollutants, unacceptable amounts of dangerous and ecologically damaging chemicals have been and continue to be discharged into the environment. The inadequacy of so called "end-of-pipe" pollution control approaches can be discussed by reviewing the following examples of local and global environmental pollution problems [38–47], which threaten human health and the integrity of the environment:

- *Acid rain*: The sulfur and nitrogen oxides produced by coal-burning power plants, automobiles, and several factories form sulfuric and nitric acids in the atmosphere. The acidic rain, snow and dust washed back to the earth is driving the pH of water and soil, in poorly buffered regions, to levels so low that they cannot support most forms of life. In fact, many aquatic environments have become toxic to fauna, flora and aquatic organisms. Acid precipitation also interacts with other wastes. Research has found that [40, 42, 43]: (a) half the streams of the mid-Atlantic and southeastern states are either now acidic or are on the verge of acidity; (b) 25% of the nitrogen entering the Chesapeake Bay is provided by acid deposition and has contributed to the unusual algal blooms which are driving the dissolved oxygen levels down and threatening the regional fisheries industry, and; (c) trees in National Forests have been damaged by a combination of ozone, acid rain, acid fog, and consequently suffer increased susceptibility to infectious agents.
- *Ozone hole*: Since a large hole in the stratospheric ozone envelope was discovered over the Antarctic, it has been established that the thinning of the

ultra violet-filtering ozone layer is a general global phenomenon and poses serious risk of increased skin cancer cases, severe sunburn, and damage to the eyes [38, 41]. In addition, UV light causes direct damage to plants, and slows the rate of photosynthesis in oceanic phytoplankton that produces about 80% of the Earth's new oxygen.

- *Climate change*: There has been growing evidence that the Earth's temperature is rising due to an increased rate of buildup of greenhouse gases (including carbon dioxide, chlorofluorocarbons, and methane in the upper atmosphere). Predictions of the effects of this global warming include elevation of sea level around the world and severe disruption of weather patterns in many regions. Evidence of a long-term warming trend is documented but a consensus is growing among national and international scientists, engineers and government leaders that significant global warming is occurring [40, 43, 48, 49].
- *Organic chemical wastes*: Polychlorinated biphenyls (PCBs), along with other chlorinated organic compounds, can be found sequestered in human and animal fat deposits. Though little is known about the long-term health effects of chronic exposure to low levels of these toxins, there is ample reason for concern about latent cancer, birth defects, and reproductive problems [1, 2, 41]. Over 32% of the PCBs produced have already been lost to the environment. Two-thirds of these are in landfills, while the rest are scattered in various environmental media (the air, lakes, rivers, oceans, bottom sediments). PCBs have been entering complex environmental cycles and undergoing biological magnification in fish, fish-eating birds, and aquatic mammals. There is growing concern about the production of dioxins during incinerator operation. Much of the opposition to the siting of waste incinerators is based on fear of the release of dioxins and dioxin-like substances (such as chlorodibenzofurans from PCBs) that are produced during combustion of various halogenated organic compounds.
- *Heavy metals wastes*: Heavy metals pose a hazard to humans and many organisms due to their toxicity and tendency to accumulate within the body. Like halogenated organics, some of them also undergo biological magnification as they are processed through plants and animals at successive levels of the natural food chains/webs [1, 2, 16, 17]. Mercury, cadmium, chromium, zinc, arsenic, and lead are among the more toxic inorganics found in the environment, in hazardous waste dumps and existing industrial hazardous waste streams. Generally, 20% of the hazardous wastes generated contain acids carrying these dissolved metals. Major sources include metal plating, forging, photo-processing, mining, and electronics industries.
- *Hazardous waste dumping*: The impact of decades of land disposal of hazardous waste is inestimable. The U.S. EPA reports that 30,000 to 50,000 hazardous waste sites exist. 90% of those in the eastern half of the country are leaking toxic substances into groundwater. According to the Office of Technology Assessment, the number of sites eligible for federal and state Superfund cleanup assistance is expected to grow to 10,000 at an estimated

remediation cost of $100 billion over the next 40 years. Over 950 sites have been placed on the National Priorities List because of threats to nearby populations via air, surface water or groundwater [1, 2, 16, 17].

2.2 Waste Management Hierarchy

The U.S. EPA has already established a hazardous waste management hierarchy [20, 21]. This hierarchy represents an attempt to build environmental protection into the industrial waste management process. It encourages industries to reduce their hazardous wastes at the source and to recycle rather than treat and/or dispose of wastes to land, air and water. This hierarchy of hazardous waste management is presented in Fig. 2. The following is a summary.

2.2.1 Waste Minimization

Figure 2 indicates that the top three sections of the pyramid include source reduction at the top, onsite recycling next and offsite recycling in third order of preference. Source reduction and recycling taken together represent two approaches to waste minimization.

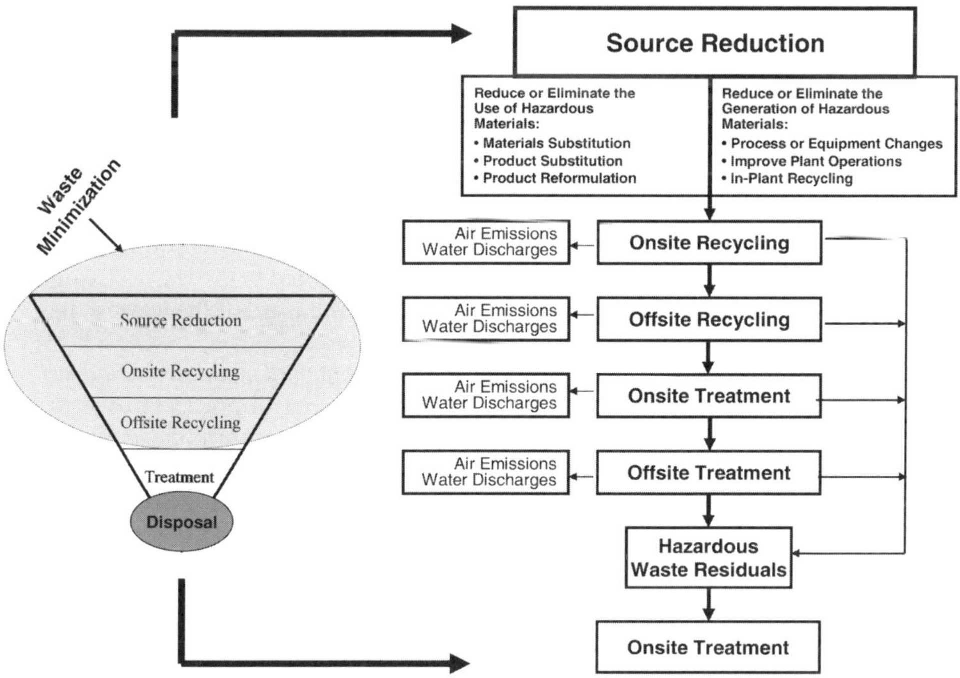

Fig. 2 The hazardous waste management hierarchy

Another term, "waste reduction", is sometimes used interchangeably with "waste minimization". However, as generally used, waste reduction includes treatment along with source reduction and recycling. Waste minimization effectively reduces the amount of hazardous material that permanently leaves the production process as waste. It was the intention of the EPA that industry implements source reduction and recycling before resorting to treatment and/or disposal [20, 21]. However, until recently, this intent has not been widely implemented.

2.2.2 Source Reduction

Source reduction is pollution prevention, which reduces the toxicity and/or quantity of hazardous material at the source. Therefore, it is fundamentally different from the pollution management practices of recycling, treatment and disposal. Because waste minimization and source reduction have been defined in various ways, the following are considered not to be source reduction [16, 17, 20–23, 26, 36, 44]:

- *Actions taken after a hazardous waste is generated.* This clearly excludes any form of treatment (including detoxification, incineration, thermal, chemical or biological decomposition, stabilization through solidification, and embedding or encapsulation).
- *Recycling.* If, however, residual hazardous material contained within an existing process or operation is fed back into the process, then it is considered to be source reduction.
- *Actions concentrating the constituents of a hazardous waste to reduce its volume.* Volume reduction operations, if they simply reduce the total volume without reducing the toxicity of a waste, do not qualify as source reduction.
- *Actions diluting the hazardous waste after generation to reduce its hazardous characteristics.* Dilution is the opposite of volume reduction. As with volume reduction, waste dilution constitutes treatment applied to the waste stream after generation. It does not reduce the amount or hazard of the waste at its source of production.
- *Actions shifting hazardous waste from one environmental medium to another.* Many of the waste management, treatment and control practices used to date have simply collected pollutants and moved them from one environmental medium to another, as follows: (a) collection of pollutants from air and water via pollution control devices, which are then legally disposed of in land disposal sites, and (b) transfer of volatile pollutants from surface impoundments, landfills, water treatment units, groundwater air stripping operations, to the air through evaporation.

2.2.3 Recycling

Recycling is the use, reuse or reclamation of wastes after they have been generated. Through recycling and reuse, hazardous wastes are routed into production processes rather than being released to the environment. Reclamation is usually considered to be a part of recycling because it recovers raw material for reuse. The processes used to reclaim useful materials from waste often generate hazardous wastes of their own [16, 17, 20–23].

The recycling option is less preferable and should be resorted to after all feasible source reduction options have been explored and implemented. Onsite recycling has a higher priority than offsite recycling because reducing transportation, storage and other handling of hazardous wastes reduces the risks to health and the environment.

2.3 Incentives and Strategies

All waste generators have powerful incentives to reduce the volume or hazardous nature of their wastes. These incentives may be economic, regulatory, or a variety of other incentives either unique to the individual industrial firm or common to business in general [14, 20–23, 25].

The overall economic incentive is the potential for increased competitive advantage through lower operating and production costs. Economic incentives apply to the generation of both non-hazardous and hazardous wastes. Such incentives can be grouped into four basic categories: reduced waste management costs, improved operations, reduced liability risks, and increased competitive advantage. Compliance with federal and state regulatory requirements is a key incentive for waste minimization. Other incentives include: positive public image, healthy environment, product quality improvement, government assistance programs, and policy adherence. These incentives are more difficult to quantify than either regulatory or economic incentives, though no one should doubt or underestimate their value [20–23].

As indicated previously, source reduction is pollution prevention in contrast to practices that manage hazardous waste after it has been generated. As such, source reduction requires a fundamentally different approach. Successful source reduction depends partly on technological change. Source reduction techniques vary greatly from one industry to another. Approaches to source reduction are in four major categories that follow the techniques presented in Fig. 3. They include the following [1, 14, 20–23, 25]:

- *Good operating practices:* This includes many procedural changes that can be implemented in many areas of industrial plant operation. Many effective operating practices can be implemented at low cost giving a good return on the investment in a relatively short time. The main specific aspects of good operating practices are: waste management cost-accounting, inventory man-

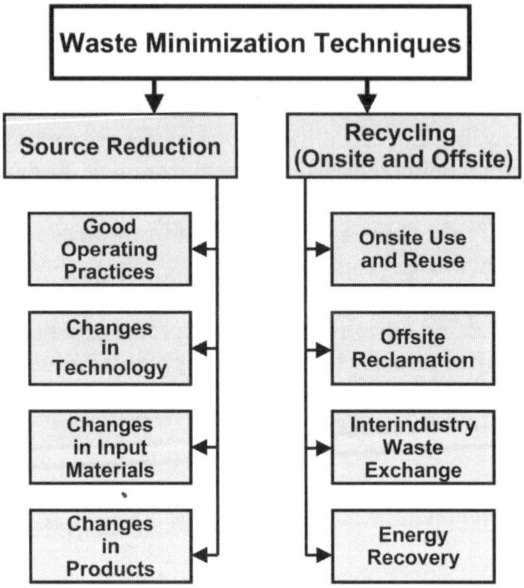

Fig. 3 Waste minimization techniques

agement, procedural scheduling, material handling improvements and loss prevention, waste stream segregation, and personnel education, communication and involvement.

- *Changes in technology:* Source reduction strategies involving changes in industrial technology tend to be industry- and process-specific and require a good understanding of the details of a facility's operation. Categories of technological changes to achieve source reduction include: process changes, equipment, piping or layout changes, and additional automation.
- *Changes in input material:* This is an active and growing area of industrial research and development. Less toxic substitute products are being developed and tested. Procedures for increasing the purity of raw materials include: material purification and material substitution.
- *Changes in product:* The amount or toxicity of hazardous waste generated from a product's end-use may be reduced by substitution of one product for another, conservation of the product, or altering the composition of the product.

2.4 Recycling Techniques

Recycling is the use, reuse, or reclamation of all or part of a waste. It is the most preferable waste minimization method following source reduction [16, 17, 20–23]. The most preferable form of recycling is that which requires the least amount of management and returns the greatest proportion of the material for reuse. It is economically advantageous because the need for raw materials is reduced. It is environmentally beneficial because it conserves natural resources and avoids the higher risk that comes with treatment and disposal operations. Recycling generally falls into two main categories (Fig. 3):

- *Onsite recycling:* It is the reuse of waste materials at the site of generation. It may be used in the same or another process (like the reuse of cleaning washes and of solvents in the production process). This type of recycling reduces the cost of raw materials as well as the cost of waste disposal.
- *Offsite recycling:* It involves transporting the waste to a commercial recycler which processes and returns the material to the generator (such as in solvent leasing, battery recycling, used oil recycling). Offsite recycling generally reduces disposal costs, resulting in some savings on purchase of these materials.

The suitability of a waste for recycling depends mainly on the purity, concentration, and chemical form of the reusable material in the waste. For this reason, some waste materials must be treated before being recycled. This "treat/purify-then-recycle" option is less desirable than direct in-process recycling/reuse without treatment [20–23, 25, 50], as follows:

- *Onsite Use and Reuse:* In-process and other onsite recycling options may be made more feasible by the adoption of other source reduction practices, such as waste segregation to cut down contamination of materials (e.g., solvents, baghouse dusts, lubricating oils, or rinse waters). Reuse as raw material substitute for another process.
- *Offsite Reclamation:* Reclamation is the recovery of a valuable material from a hazardous waste. Its techniques differ from use and reuse techniques in that the recovered material is sold to another entity rather than being used in the facility generating it. This option should follow exhaustion of onsite recovery options because of the increase in potential liability, both economic and environmental, with the transportation of waste to the offsite facility.
- *Inter-industry waste exchange*: This approach involves the transfer of treated waste and untreated hazardous wastes to another industrial user (exchange for use as raw material). If the industry can find another industry in need of the waste material, it can eliminate disposal costs for that waste stream and may recover some of the costs for the raw materials used in the process.
- *Energy recovery*: When waste is used for energy production, it can be processed in a variety of ways (like in cement kilns, asphalt plants, co-firing fossil-fuel-tired plants, incinerators equipped with an energy recovery system). Larger firms with their own power generation facilities may have the flexibility to co-fire moderate to high BTU wastes [16, 17]. Waste characteristics may restrict use of the waste as a fuel due to other regulatory requirements (such as public health and environmental risks) or concern for boiler corrosion.

Commercial fuels-blending programs are usually run in conjunction with treatment or incineration firms. These firms blend a variety of high BTU wastes with varying compositions, from a number of generators to produce a fuel with particular specifications. These blended fuels are usually used in cement kilns or asphalt plants. Waste generators considering such alternatives should perform site inspections and investigate the blending firm and the ultimate user of the fuels. Full regulatory and legal reviews should also be conducted.

Incineration is another form of energy recovery [1]. Some hazardous waste incinerators and many refuse incinerators are equipped with energy recovery equipment. While some major facilities operate either or both types of incinerators, most firms must resort to offsite treatment. These treatment facilities should be carefully evaluated from an economic, regulatory, and operations standpoint. While these facilities reduce volume and, generally, toxicity, air pollutant generation and the final incinerator ash disposal must still be considered and may impact a generator's liability.

2.5 Barriers

The shift of emphasis from "end-of-pipe" waste management and pollution control to waste minimization is not without its obstacles [16, 17, 20–23]. However, the barriers to waste minimization are similar to those of any other major change in business practices. The recognition of the hazards, liabilities, and costs associated with hazardous waste generation is relatively new to the industrial community and there are significant limitations on available financial resources, information, and technical capabilities. Waste minimization can be further limited by institutional barriers due to the persistent "end-of-pipe" outlook on hazardous wastes that is common to both government and industry. Some of the barriers to waste minimization include [16, 20–23, 28–30, 36, 37]:

- *Financial*: Many waste generators lack the available capital to implement costly equipment changes and process modifications. However, many of the opportunities to minimize waste (both source reduction and recycling) require little or no significant capital outlay or equipment changes. Management personnel should also be encouraged to adopt a thorough waste management cost-accounting system to assess the feasibility of waste minimization for their facility. An improper accounting system may serve as a disincentive to adopting a waste minimization program.
- *Informational*: Generators may lack knowledge of waste minimization methodologies and regulatory requirements, or lack the technical expertise to institute waste minimization strategies. The waste minimization options do not usually involve highly technical information. Providing locally appropriate information in an accurate and understandable manner is essential to the widespread adoption of waste minimization strategies. At a minimum, one should be sufficiently knowledgeable to respond to inquiries and to identity sources of further information.

- *Technical*: Technology, in general, is not a factor that should seriously limit a generator from implementing many of the methods of waste minimization. However, in some cases, the lack of specially designed equipment and technical expertise do impede the changeover to practices that generate less waste or recycle what is generated.
- *Institutional*: Hazardous waste generators are often unable to devote attention to assessing their operations for waste minimization opportunities because they are putting so much energy, money and time into complying with existing regulations and keeping ahead of new regulations. In many cases, they may lack confidence in the regulating community. Lack of knowledge among regulators and lack of flexibility in regulations have been cited as limitations. Existing regulations and market factors may work against efforts to implement waste minimization. While regulations limit the amount of hazardous waste that they can accumulate, the recyclers often require large quantities of waste from clients to qualify for their services. This is compounded by the fact that fluctuations in market values of virgin stock can cause unpredictable changes in demand for recycled raw materials.

2.6 Case Study

Waste minimization has been proven to be an effective and beneficial operating procedure, for instance, in drilling operations [51–62]. Currently, there are many economically and technically feasible techniques that can be used. It is important to consider all costs, including waste management and disposal costs, when evaluating the feasibility of a waste minimization option [16, 17]. In fact, many oil and gas companies have implemented waste minimization techniques and have gained benefits (for instance, reduced operating and waste management costs, increased revenue, reduced regulatory compliance concerns, reduced potential liability concerns, and improved company image and public relations [56–59]). Choosing feasible source reduction and recycling options (i.e., a waste minimization program) is a smart decision. The following section provides a general overview of two main waste minimization techniques for wastes arising from oil and gas drilling operations [63–100]:

2.6.1 Source Reduction

The various source reduction opportunities for wastes arising from a drilling operation include:

2.6.1.1 Preplanning

The best place to start waste minimization efforts for a drilling operation is in the planning stages. The drilling plan should be evaluated for potential waste generation and modified to take advantage of the other source reduction and recycling options discussed below. This preplanning can make a significant impact on the waste management requirements of the drilling operation, as follows:

- Drill site construction and rigging-up: A preplanning opportunity for a drilling operation is in the construction of the location and roads. The drilling location and the associated roads should be planned so that they are constructed such that stormwater runoff is diverted away from the location, and that the location's stormwater runoff, which may be contaminated, is collected. Construction of the location and roads should be planned so that erosion is minimized. These steps will help minimize the volume of contaminated stormwater runoff to be managed [59, 63–66].
- *Drilling fluid systems*: It is important to design the drilling fluid system with specific waste minimization practices in place. These waste minimization techniques should be integrated into the drilling fluid system portion of the drilling plan [51, 54, 55, 67–71].
- *Pit design*: Another consideration in preplanning for a drilling operation is the design of the reserve pit. A V-shape design, for example, will provide advantages with respect to waste generation and operational costs. The open end of the "V" faces the drilling rig and the cross-sectional view looks like a squared-off funnel. This V-shape design prevents mud from channeling from the discharge point to the suction point, as it must travel the full length of the pit. Also, because the V-shaped pit is long and narrow, it is easier to construct and line [63, 65, 72].

2.6.1.2 Product Substitution

Product substitution is one of the easiest and most effective source reduction opportunities. A few examples of effective and beneficial product substitution for drilling operations are listed below [63–66, 73–78]:

- *Drilling fluids*: Many companies have found that the substitution of low toxicity glycols, synthetic hydrocarbons, polymers, and esters for conventional oil-based drilling fluids is an effective drilling practice. The use of substitute drilling fluids eliminates the generation of oil-contaminated cuttings and other contamination by the oil-based fluid (such as the reserve pit and accidental releases). Drill site closure concerns are also reduced. Drilling engineers have published numerous technical papers that describe the successful application of substitute drilling fluids.
- *Drilling fluid additives*: Many of the additives used in the past for drilling fluids have contained potential contaminants of concern, such as chromium in lignosulfonates. Also, barite weighting agents may contain concentrations of heavy metals such as cadmium or mercury. The use of such additives has diminished. However, it is important to select additives that are less toxic and

that will, therefore, result in less toxic drilling waste. The design of the drilling fluid system is the best place to implement this product substitution opportunity.

- *Pipe dope*: Pipe connections require the use of pipe dope. American Petroleum Institute (API) specified pipe dope contains about 30% lead by weight and, therefore, can be of concern when disposed of. One simple waste minimization technique is to ensure that all pipe dope is used and containers are completely empty. However, lead-free, biodegradable pipe dopes are now available and, if feasible, should be substituted for API specified pipe dope. Even if API specified pipe dope is necessary for making the required connections, pipe supply companies should be asked to provide pipe with lead-free pipe dope on the thread protectors.
- *Organic solvents*: Organic solvents (e.g., trichloroethylene and carbon tetrachloride) are commonly used for cleaning rig equipment and tools. These solvents, when spent, become listed hazardous oil and gas wastes and are subject to stringent regulation. Alternative cleaning agents (e.g., citrus-based cleaning compounds and steam) may be substituted for organic solvents, which will eliminate a hazardous waste stream along with the associated waste management and regulatory compliance concerns. Another solvent commonly used on drilling rigs is Varsol (also known as petroleum spirits or Stoddard solvent). While most Varsol has a flashpoint below 140°F, which is therefore a characteristically hazardous waste when spent, some suppliers may provide a "high flash point Varsol" with a flash point greater than 140°F.
- *Paints and thinners*: Although oil-based paints and organic solvents (thinners and cleaners) are used less frequently these days, they are still used. These paints and thinners provide an excellent product substitution opportunity. Water-based paints should be used whenever feasible. The use of water-based paints eliminates the need for organic thinners (like toluene). Organic thinners used to clean painting equipment are typically listed as hazardous waste when spent. This substitution can eliminate a hazardous waste stream and reduce waste management costs and regulatory compliance concerns.

2.6.1.3
Equipment Modifications

The drilling rig's diesel power plants typically generate large volumes of waste lubricating oil and lubricating oil filters, which contain high concentrations of aliphatic and polycyclic aromatic hydrocarbons [1, 2]. A lube oil testing program combined with extended operating intervals between changes is effective. However, an equipment modification can also effectively reduce the volume of waste lubricating oil and filters. Generally, commercial vendors offer a device called a lube oil purification unit, which removes particles greater than 1 micron in size and any fuel, coolant, or acids that may have accumulated in the oil [63]. The unit does not affect the functional additives of the lube oil. The lube oil is circulated out of the system and through the purifier. The purified lube oil is then returned to the engine's lube oil system. Many operators have found that use of lube oil purification units has significantly reduced the need for lube oil changes, waste lube oil management, and concurrently, the cost of replacement lube oil [79–81]. Also, a new engine that has been fitted with a lube oil purification unit will break in better and operate more efficiently over time.

2.6.1.4
Process or Procedural Modifications

These modifications include the following [82–89]:

- *Slim holes*: The drilling industry has improved the technology of "slim hole" drilling over the past few years. "Slim hole" drilling should be considered when planning any drilling project. If used, slim hole drilling reduces the volume of waste drilling fluid and the volume of drill cuttings.
- *Solids control for the drilling fluid system*: An effective way to reduce the volume of drilling fluid waste is the use of solids control. The efficient use of solids control equipment (for example hydrocyclones and centrifuges) in combination with chemical flocculants minimizes the need for makeup water to dilute the fluid system. An enhanced solids control system designed to compliment a specific drilling operation is a very effective waste minimization technique.
- *Material balance and mud system monitoring*: Companies have found that diligent and comprehensive monitoring of drilling fluid properties is effective in reducing the frequency of water and additive additions to the system. Such a system is also referred to as "Integrated Drilling Fluids Management". Monitoring devices at various points in the system facilitate the immediate identification of unwanted changes in the drilling fluid system and allow the necessary corrections to be made. This technique, in addition to the solids control described above, can significantly reduce the costs of the drilling fluid system and the volume of drilling waste remaining at the end of the operation.
- *Closed-loop drilling fluid systems*: Closed-loop drilling fluid systems provide many advantages over conventional earthen reserve pits. Closed-loop drilling fluid systems use a series of steel tanks that contain all drilling fluid and equipment used to remove cuttings. These systems enhance the ability to monitor fluid levels and characteristics. The result is more efficient use of the drilling fluid and less drilling waste remaining at the end of the operation.
- *Mud runoff from pulled drill string*: Running drill pipe into and out of the hole can contribute to the volume of waste in the reserve pit. Lost drilling mud and the excess rig-wash required for cleaning it from the rig floor can be major contributors, and can be minimized. Devices are currently available that wipe clean the inner diameter of the drill pipe as it is pulled so that the mud does not run onto the rig floor. Therefore, drilling mud losses and the need for rig-wash are reduced.

2.6.1.5
Reduction in Water Use

Currently, there are many techniques which help reduce the amount of water use in drilling operations [90–94], as follows:

- *Rig-wash hoses*: A simple way to minimize the volume of waste rig-wash is to use high-pressure/low-volume nozzles on the rig-wash hose. A rig-wash hose left running can contribute significantly to the volume of waste in a reserve pit and the water needs for the drilling operation. If feasible, collection and treatment of rig-wash for reuse is a good waste minimization technique.
- *Drilling fluid systems*: Improved design and operation of drilling fluid systems can also reduce the need for water. Waste minimization opportunities (such as solids control and detailed system monitoring) have been proven effective in reducing the amount of makeup water needed in a drilling operation.
- *Dewatering waste drilling fluids*: Water can be reclaimed from waste drilling fluids by using mechanical or chemical separation techniques. Large bowl centrifuges, hydrocyclones, and/or chemical flocculants may be used to dewater waste drilling fluids. The reclaimed water may then be reused, reducing the demand on new water sources. Proper application of dewatering can result in a reduction in the volume of drilling waste to be managed, saving waste management costs, easing site closure concerns and costs, and reducing future potential liability concerns.

2.6.1.6
Housekeeping and Preventative Maintenance

An important source reduction technique is preventive maintenance. This includes the following [63–66, 95–100]:

- *Drill site construction and rigging-up*: This involves the use of heavy equipment, which should be well maintained to reduce the potential for fuel and lubricating oil leaks that may contaminate the site. Preventative maintenance during the construction phase can help prevent the generation of contaminated soil and water. For example, secondary containment beneath fuel storage drums can prevent releases to soil and water.
- *Various types of containment*: A good housekeeping practice, which helps reduce the amount of soil and water contamination, is to install containment devices. Containment devices can capture valuable released chemicals that can be recovered and used. Some examples of containment include: drip pans beneath lubricating oil systems on engines; containment vessels beneath fuel and chemical storage tanks/containers; and drip pans beneath the drum and container storage area. Numerous industrial facilities have implemented good preventive maintenance programs to reduce the amount of crude oil, chemicals, products, and wastes that reach the soil or water. In general, these programs have been found to be cost effective in the long term.

- *Preventive maintenance*: The companion of good housekeeping is preventive maintenance. Regularly scheduled preventive maintenance on equipment, pumps, piping systems and valves, and engines will minimize the occurrence of leaks and releases of chemicals to containment systems. Preventive maintenance programs have been implemented and found to be quite successful. The programs have resulted in more efficient operations, reduced regulatory compliance concerns, reduced waste management costs, and reduced soil and/or ground water clean-up costs.
- *Chemical and material storage*: Another important aspect of good housekeeping is the proper storage of chemicals and materials. Chemicals and materials should be stored such that they are not in contact with the ground. Material Safety Data Sheets (MSDSs) and other manufacturer information should be kept on file for all stored chemicals and materials. The use of bulk storage is a preferable way to store chemicals and materials. Proper storage and labeling of containers allows quick and easy identification and classification of chemicals or materials released in the event of a leak or rupture.

2.6.2 Recycling

The next preferred waste management option in drilling operations is recycling. Recycling is becoming a big business and more recycling options are available every day. The following paragraphs discuss various recycling options for drilling wastes [51, 52, 56, 59, 60, 69–71, 88].

- *Drilling fluids*: Drilling fluids comprise the largest waste stream associated with a drilling operation. The cost of closing a drilling site is increased if waste drilling fluid in a reserve pit must be dewatered and/or stabilized prior to closure. A better alternative is to recycle or reuse the waste drilling fluid. If feasible, reuse the waste drilling fluid in another drilling project. A multi-well drilling project can be designed where the same drilling fluid is used for drilling each successive well. The result can be significant cost savings and greatly reduced waste management concerns. If reuse is not feasible, waste drilling fluids can first be reconditioned and then reused. Another cost effective alternative for reuse of waste drilling fluid is in the plugging of other wells.
- *Reserve pit water*: A drilling operation should consider reclaiming water from the reserve pit using a dewatering technique. The reclaimed water can then be used as rig-wash water, makeup water for the drilling fluid system, and other rig water usage. Additionally, collected stormwater runoff may be suitable for use.
- *Paint solvent reuse*: A simple technique for reducing the volume of organic paint solvents is to reuse it in stages. An organic solvent may be used for cleaning painting equipment, but eventually it will become spent and ineffective. The "spent" solvent is not a waste if it is used for another intended

purpose. A solvent spent from cleaning painting equipment is still suitable for use in thinning paint. This simple technique can greatly reduce the volume of waste paint solvent that may be subject to stringent hazardous waste regulation.

- *Commercial chemical products*: In general, procedures that recycle any unused chemical products should be implemented. Whenever a vendor is contracted to supply chemicals, the vendor should be required to take contractual responsibility for unused chemical products and the containers in which they were delivered. As discussed under the source reduction opportunity, commercial chemical products that are returned for reclamation or recycling are not solid wastes. Therefore, managing chemical products properly can avoid the unnecessary generation of unused chemicals that must be disposed of. In many instances, those chemical wastes would be hazardous and subject to stringent regulation.

3 Molecular Nanotechnology

The relationship between new technologies and the environment is very complex. Various human technologies have caused a tremendous harm to the environment [38–40]. New technologies are often cleaner and safer than the older technologies they replace [31–33]. The following is a summary.

3.1 Theory

Nanotechnology is an anticipated manufacturing technology giving thorough, inexpensive control of the structure of matter. The term has sometimes been used to refer to any technique able to work at a submicron scale, sometimes called molecular nanotechnology (MNT). The central theory of nanotechnology is that almost any chemically stable structure that can be specified can in fact be built [31–33].

An "assembler" is a device having a submicroscopic robotic arm under computer control. It is capable of holding and positioning reactive compounds in order to control the precise location at which chemical reactions take place. This general approach allows the construction of large atomically precise objects by a sequence of precisely controlled chemical reactions, building objects molecule by molecule [31]. If designed to do so, assemblers will be able to replicate (i.e., build copies of themselves), and hence be inexpensive. By working in large teams, assemblers and more specialized nano-machines will be able to build objects cheaply [32–33]. By ensuring that each atom is properly placed, they will manufacture products of high quality and reliability. Left-over molecules would be subject to this strict control as well, making the manufacturing process extremely clean [101, 102].

Traditional industrial technologies operate from the "top-down". Blocks of raw materials are cast, or machined into precisely formed products by removing unwanted matter. Results of such processes may be rather small (e.g., integrated circuits) or very large (e.g., ocean liners). However, in all cases matter is being processed in pieces far larger than molecular scale [46, 103–109]. This "top-down" bulk technology (TDBT) is certainly capable of yielding products of fairly high precision and complexity.

Rather than being produced through large amounts of material being sawed, planed, and ground to form, most such objects created via nanotechnology are constructed by tiny molecular machines (such as cells and organelles) working from the "bottom-up". By organizing individual atoms and molecules into particular configurations, these molecular machines are able to create works of astonishing complexity and size [110–112]. This approach can produce results that would seem impossible if judged by the standards of conventional "top-down" production technology.

The plausibility of this approach can be illustrated by the "ribosomes", which manufacture all of the proteins used in all living things on this planet. A typical ribosome is relatively small (a few thousand cubic nanometers, "nm"), and is capable of building any protein by stringing together amino acids (the building blocks of proteins) in a precise linear sequence [113–117]. To do this, the ribosome has a means of: (a) selectively grasping a specific transfer RNA (which in turn is chemically bonded by a specific enzyme to a specific amino acid), (b) grasping the growing polypeptide, and (c) causing the specific amino acid to react with and be added to the end of the polypeptide. The instructions that the ribosome follows in building a protein are provided by mRNA (messenger RNA). This is a polymer formed from the four bases adenine, cytosine, guanine, and uracil. There is a sequence of several hundred to a few thousand such bases codes for a specific protein. The ribosome reads this control tape sequentially, and acts on the directions it provides [115, 116].

The molecular machinery responsible for this amazing achievement of production is capable of such dramatic results because it performs operations in parallel and from the "bottom-up" [118, 119]. Putting these natural molecular machines to work is nothing new; what makes nanotechnology different is that it involves the attempt to go further than natural mechanisms permit. Using special bacterium-sized assembler devices, nanotechnology would permit exact control of molecular structures that are not readily manipulable by organic means on a programmable basis [120, 121].

3.2 Environmental Benefits

It is believed that the environmental benefits of nanotechnology would be diverse and dramatic. Since nanotechnology involves atom-by-atom construction, it will be able to create substances without producing the harmful byproducts that most current manufacturing processes produce. Nano-devices will operate in a liquid containing the necessary raw materials (usually carbon or silicon) and will simply plug the appropriate atoms in the appropriate places to produce the desired end product [110–112, 114, 118, 119, 122–125]. Such processes should produce few byproducts, and those byproducts can be readily purified (by other nano-devices) and recycled back into feed-stocks.

Furthermore, most products of nanotechnology will be made of simple and abundant elements. Carbon (as in diamond or diamondoid form) is seen as the basis of most nano-manufacturing. Products made of such materials will be very strong, meaning that smaller amounts of material can be used, and carbon is an abundant material, meaning that little in the way of exploration and extraction will be needed. As a greenhouse remediation measure, nano-devices could even extract carbon dioxide from the air if desired.

This environmentally clean manufacturing is a significant benefit of nanotechnology [31–34, 126, 127]. In addition, when materials are inexpensive, and structures of great strength and low weight can be manufactured cheaply, energy requirements for many activities drop enormously.

Many scientists and engineers also believe that atom-by-atom manufacturing will make low-cost, high-efficiency solar cells practical. The combination of reduced energy demand and inexpensive solar power may make most of today's power generation and transmission infrastructure unnecessary. The payoff from such improvements would be enormous, not only in terms of reduced pollution from power generation, but also in terms of reduced environmental impact all along the production and distribution chain [31, 126, 127].

Beyond these impacts, more advanced nanotechnology may allow active remediation of many environmental problems. For example, toxic wastes in contaminated aquifers may be neutralized by specially designed nano-robots (nanobots) that selectively capture undesirable molecules and then either sequester them for removal or break them down into harmless substances [114, 118, 119, 124]. While nano-devices cannot, for example, render radioactive materials non-radioactive, they could capture molecules of radioactive waste and concentrate them into a form that would be easily removed [31–33].

4
Total Environmental Sustainability

Environmental sustainability based on technology-to-date (TTD) is pessimistic [1, 26, 28–30]. Emerging molecular nanotechnologies (MNT) on all industrial fronts offer radical tools capable of providing human society with the upper hand in the struggle toward environmental sustainability and economic growth for the first time [31–33]. Furthermore, MNT will have extra benefits for human civilization beyond the remediation of many environmental disasters accumulated since the Industrial Revolution of the eighteenth century, facilitating the production of unlimited material and energy with ultra green processes. The following is a summary.

4.1
Current Human Impact

The technologies currently employed for production are top-down bulk technologies (TDBT). Due to TDBT's inherent shortcomings (low efficiency, imprecision) throughout the processes, energy and material are excessively wasted; and pollutants and wastes are extensively generated. In addition, human societies consume large quantities of energy and material, and generate significant amounts of pollutants and wastes [16,17,24,26,36,37]. These, of course, threaten the sustainability of our environment.

Human activities can be summarized into: energy production, material production, product manufacturing and consumption (Fig. 4). Because TTD is primitive in contrast to MNT, all human activities are currently exerting tremendous pressure on environmental sustainability. In addition, current pollution abatement technologies are based on TDBT; and the secondary pollution generated essentially cancels the benefit. Available data indicate that the global environment is still deteriorating [1, 2, 16, 17, 38, 40–43, 48, 49]. Humanity is suffering from degrading environmental quality, and diminishing natural

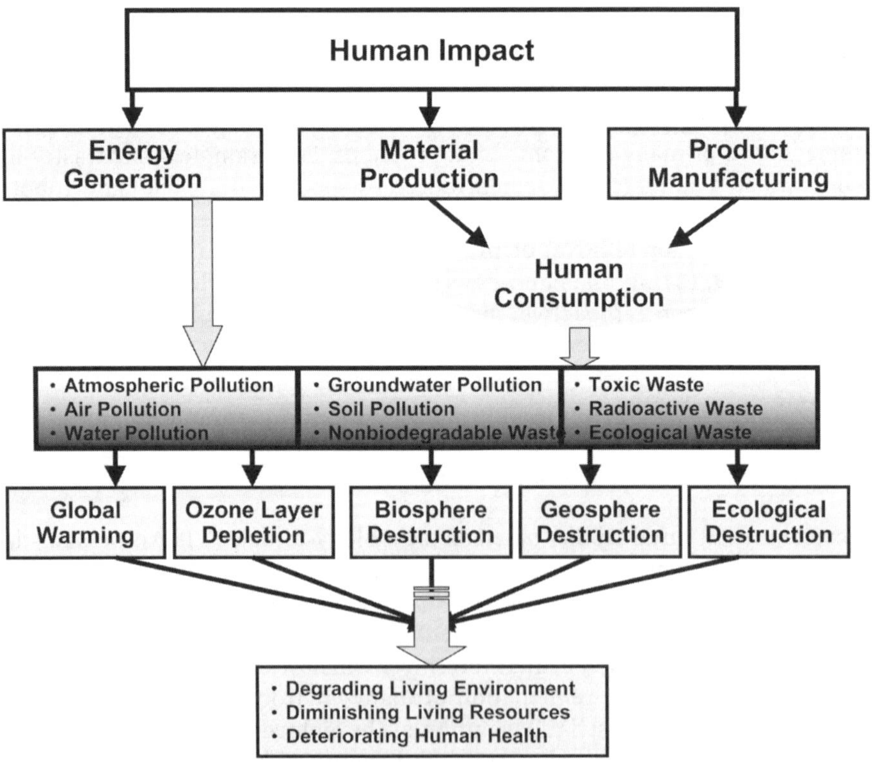

Fig. 4 Human impact on environmental sustainability

resources. If there is no fundamental technological or non-technological paradigm shift – a catastrophic environmental destruction is predicted.

While MNT is still under development, it's worth diagnosing the causes and effects of the current environmental predicament so that the targets for MNT Research and Development (R&D) into green production and products, as well as into restoring the environment, can be identified. Two examples of current human impact on environmental sustainability are discussed below [42, 128, 129]:

- *Energy production*: Major energy production processes currently adopted are hydroelectric and fossil fuels. They are the major pollution sources and destructive forces to the environment. Combined they produce more pollution and destroy more environment than any other single industry. Figure 5 summarizes the causes and effects of energy production to the destruction of the environment. Other minor power generation processes are solar, wind, geothermal, ocean thermal energy, tidal, and other renewable resources. They are the *ad hoc* energy production alternatives, in response to the serious ecological impact of the major processes or depletion concern of nonrenewable fossil fuel. Nevertheless, renewable energy facilities also affect

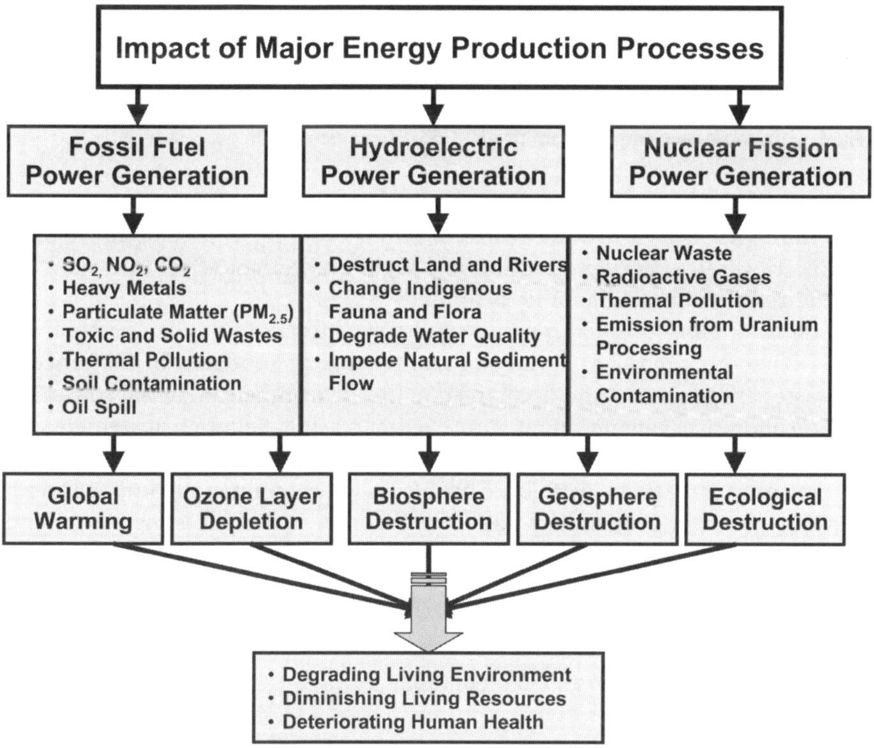

Fig. 5 Environmental impact from major energy production processes

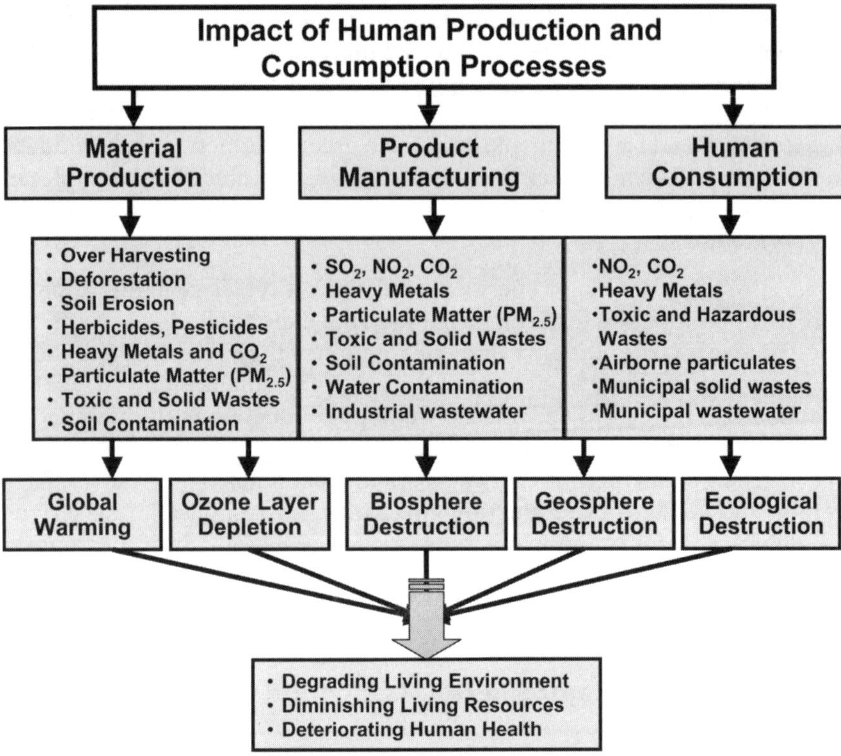

Fig. 6 Environmental impact from production and consumption processes

wildlife, involve hazardous wastes, or require cooling water, and due to their lower overall energy production and higher cost, their effect in enhancing environmental sustainability is limited.

- *Production and consumption*: For millennia, humanity has been employing TDBT to obtain raw material and manufacturing products. In the process, wastes are generated and discharged to the environment. Material consumption and waste generation rates are ever increasing. Pollution abatement and waste reduction technologies based on TDBT offer no radically solutions; they only transfer or distribute problems to less immediate impact areas, while generating secondary pollution. Figure 6 summarizes the causes and effects of environmental destruction by human production and consumption processes.

4.2 Environmental Protection and Remediation

MNT's ability to precisely manipulate atoms and molecules makes it possible to revolutionize TTD for energy and material production [31–33]. It is vital to direct MNT innovation toward environmental protection and remediation through integration, so that a sustainable, pollution-free, resource-abundant, green-wealthy nano-ecology can be obtained. In order to fulfill these goals, it is important to control pollution at source and remediate accumulated environmental problems via MNT, as follows:

4.2.1 Pollution Control at Source

Source control is most effective for TTD and MNT. TTD has been usurped to minimize human impact on the environment. Its mediocre effect is obliterated by associated secondary pollution and consumption rise. Net damage to the environment still exceeds nature's recovery capacity [31–33]. Using TDBT to meet climbing material demand rendered further environmental damage inevitable and sustainability impossible. MNT's revolutionary production processes promise eventual total elimination of all pollution sources, not to mention control, treatment or abatement.

4.2.1.1 Energy Production

Energy production processes are extremely destructive to the environment [1]. To control pollution from energy production:

- The first approach is to minimize energy consumption from the demand side. Integrating nano-electronics, nano-electromechanical systems, and nano-materials, a series of novel devices will replace all those developed by TDBT [130–132]. Lighter yet stronger material, more energy efficient and with less internal friction, created via MNT, will automatically lead to less energy consumption.
- The second approach is to develop cleaner and more efficient energy production processes from the supply side. There are 300 watts of raw solar energy irradiating each square meter of the Earth daily [31–33]. Harvesting solar energy using TDBT was proven to be uneconomical during the last energy crisis. However, results from nano-electronics and nano-energy research and development indicate that both will be technologically and economically feasible soon [131, 132].

Nano-energy developers are currently contemplating methods to reap this clean and abundant energy, as follows:

- Drexler [31–33] proposed to resurface existing roads with a coating of high-efficiency solar cells topped by a layer of tough diamond.
- Battelle researchers are developing a way to use solar energy to convert water into oxygen and hydrogen.
- AIST Japan has succeeded in using an artificial photosynthesis system to split water into hydrogen and oxygen under visible light [133].

- Scientists at CSIRO Australia are using biomimetic engineering to produce food with solar energy [132].
- Bennett's team has developed an artificial photosynthetic membrane to convert sunlight into energy [134].
- IMEC is conducting research on thin-film crystalline Si solar cells, Gas solar cells, and on new materials and technologies such as plastic solar cells to improve the efficiency and cost of solar cells [135].

From ongoing nano-energy research, it is hoped that future energy production systems can be visualized. Combusting fossil fuels and even renewable fuels for energy will become obsolete. Dams and powerhouses will be demolished and rivers will return to their original meandering courses. Energy production processes that generate CO_2, SO_2, NOx, radioactive wastes and other pollutants will be all banned. Replacing them in the energy production arena will be entirely solar-based technologies; either photoelectrolysis-based hydrogen fuel energy or photovoltaic energy systems [134, 136].

A photoelectrolysis hydrogen fuel energy system uses self-replicating nanorobots ("nanobots") to mimic the first half of the photosynthesis process to produce hydrogen gas. Fuel cells made of carbon nano-tubes could be used to interface with all energy consumers. This hydrogen fuel energy system is conceptualized as shown in Fig. 7.

Another impact from this energy production is a change in the power distribution system. High voltage transmission lines and towers not only destroy

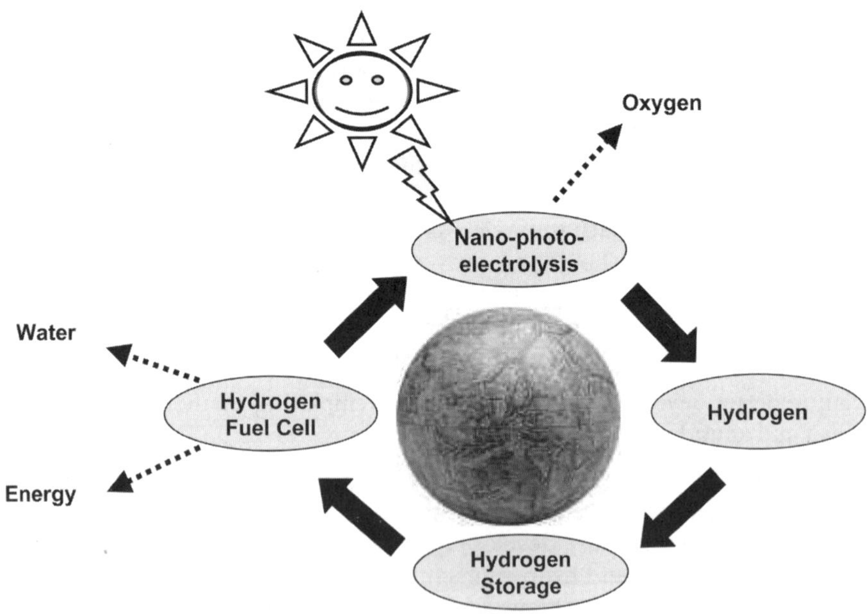

Fig. 7 MNT photoelectrolysis hydrogen fuel energy system

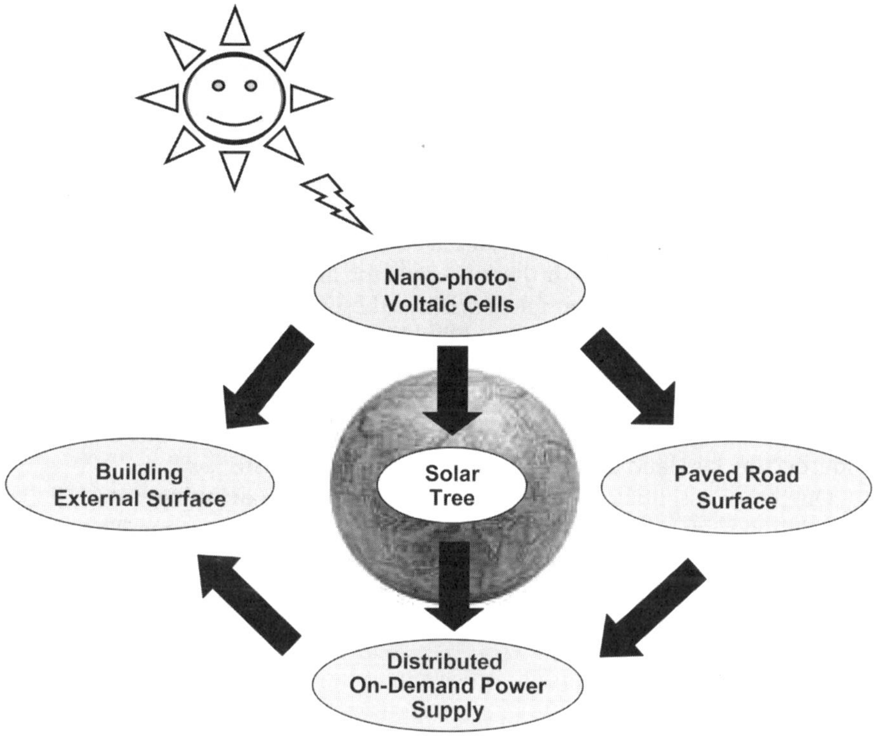

Fig. 8 MNT photovoltaic solar energy system

the environment, but their electromagnetic waves are also potentially harmful to humans. Developments (including highly advanced solar cells, inexpensive hydrogen fuel cells, and micro-generators of electricity) will make many electronic products and appliances highly mobile [131, 134, 135]. On-demand and on-location power generation will make decentralized power supplies extensive, affordable, and environmentally clean (Fig. 8).

4.2.1.2
Material Production and Product Manufacturing

Any civilization based on TDBT tends to overexploit natural resources, and generate pollution and waste during material processing and product manufacturing. MNT will render that obsolete. MNT processes will be pollution-free, generate fewer wastes and recycle 100% of raw materials [32], controlling waste and pollution. In addition, environmental remediation can proceed by nature and MNT.

MNT self-assemblers will manufacture materials with high performance, unique properties and functions that traditional industrial processes find it impossible to create [101]. Therefore, all nature exploitation activities will cease, usage of petrochemical-based material will terminate, and environmentally destructive mining will also stop.

Nano-photosynthesis can produce sugar and starch for food; and further synthesis of cellulose can produce paper and wood to avoid clear-cutting forests. Carbon retrieved from the atmosphere and recycled from existing wastes by MNT will be used to make carbon nano-tubes, with superior properties to steel. Carbon will be the most common structural and functional element for a MNT-based civilization [32, 33]. A carbon-based MNT material production model is conceptualized as in Fig. 9. If there is a specific need for metal, a nano-factory with trillions of nano-assemblers will synthesize steel, copper, and alloys in order to skip mining and refining [32, 33]. Therefore, industrial wastewater, hazardous wastes and air pollution will all vanish.

Nevertheless, to preserve quality and style of living, some farming activities will persist. Highly efficient MNT farming would use no pesticide and herbicide, occupy less land and generate no agricultural waste.

One other essential material for human survival is drinking water. Global population is increasing while fresh water supplies are decreasing. The United Nations predicts that by the year 2025, 48 countries will be short of fresh water, accounting for 32% of the world's population [31–33]. Ecological recovery via MNT will make raw water for the water supply cleaner and more abundant;

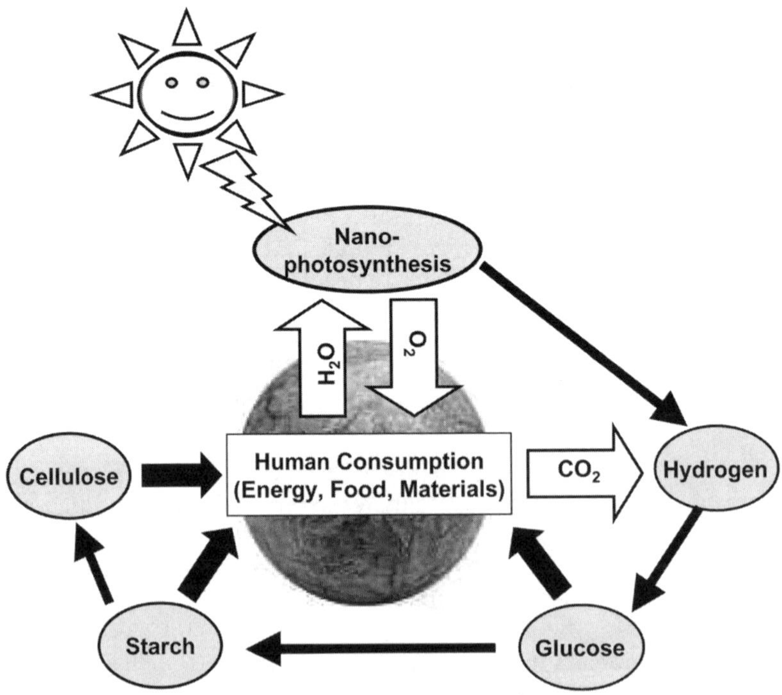

Fig. 9 MNT carbon-based material production system

however drinking water treatment will still be necessary. In general, TDBT potable water treatment consumes large quantities of chemicals for coagulation, flocculation and disinfection. The process also needs to dam the river, and it produces chemical sludge that is harmful to the environment. The treated water contains disinfectant residuals and in some cases trihalomethanes (THMs) that are detrimental to human health. The costly water distribution network not only encroaches on the environment but also provides opportunities for chemical and biological recontamination.

A series of nano-devices could be devised to revolutionize the water treatment process. Nanobots like nano-flocculant or nano-coagulant could be devised to neutralize the surface charge of suspended solids. They would be non-chemical and 100% reusable. Smart non-fouling nano-membranes or nano-separators could be developed to selectively separate dissolved solids while keeping beneficial minerals in the water, or to desalinate brine water [102]. Nano-disinfectant (such as UV) nanobots could accomplish germicidal tasks without leaving toxic residuals, and they would produce no THMs. Nano-condensers could be developed to extract water from the air. By integrating these unit nano-processes, and powered by abundant nano-solar energy, different water treatment systems may be designed to fit specific geographical conditions. Such developments would make huge waterworks with messy piping systems obsolete. On-demand and on-location generation of drinking water from liquid or vapor would make decentralized water supplies extensive, affordable, and environmentally clean. A future MNT-based system for on-demand and on-location generation of drinking water is conceptualized as in Fig. 10.

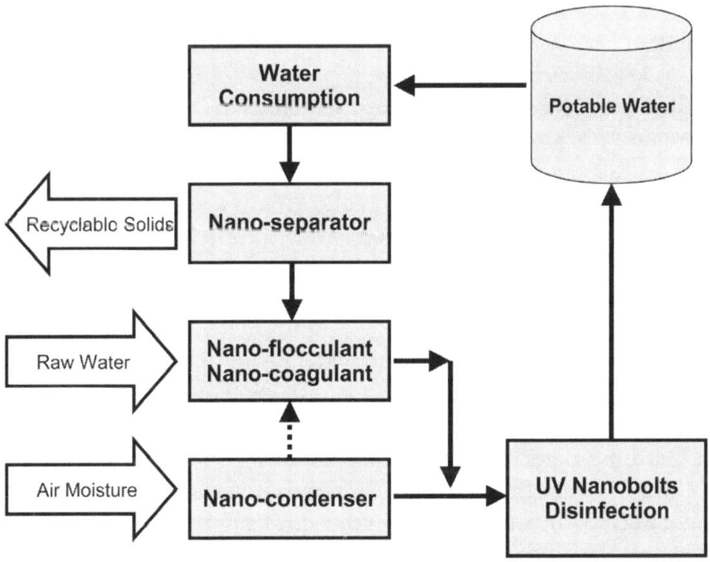

Fig. 10 MNT on-demand and on-location generation of drinking water system

4.2.1.3
Human Consumption

Various pollution sources from human consumption have significantly contributed to the destruction of the environment. Using MNT, consumer products and their packaging will be 100% biodegradable or recyclable to terminate the solid waste problem. Products (during or after consumption) that create pollution could be replaced with environmental friendly nano-material. Chlorofluorocarbon (CFC) propellants could be replaced by non-halogen gas, and volatile organic carbon-emitting solvents could use nano-water [32] as an alternative. TTD is making great strides in regards to this issue; yet with MNT, it will become a technological and economical surety. This should eventually halt air pollution originating from our consumption processes.

MNT cannot change municipal wastewater generation; however, it could revolutionize wastewater treatment [33]. Wastewater will be 100% human waste, containing no heavy metals or toxic chemicals. Human waste contains organics that are useful to nature; yet current treatment processes use electricity generated from fossil fuels to "waste" this resource. Future MNT wastewater treatment should discard the large-scale centralized wastewater treatment plant philosophy. Decentralized on-site treatment should eliminate the environmental impact from odorous and costly sewage treatment and collection systems. MNT could be used to develop smart nanobots to separate water and solids from each household or small community. The separated water could be recycled and further treated as discussed above for drinking water. Extracted solids could be stabilized biologically using nanobots, and then consumed as fertilizer or animal feed.

4.2.2
Remediation of Existing Problems

Since the Industrial Revolution, many potentially harmful environmental problems (like acid rain, global warming, ozone layer depletion, and contaminated water and soil [31–33]) have accumulated. MNT will offer the potential to return nature to its pristine state. The following is a summary.

4.2.2.1
Acid Rain and Smog

Sulfur dioxide (SO_2) and nitrogen oxides (NOx) are the primary causes of acid rain and smog [40, 43]. Acid rain causes acidification of the aquatic environment and contributes to forest and soil damage. In addition, acid rain accelerates the decay of building materials and paints, including irreplaceable buildings and statues [42]. The strength of the effect depends inversely on the buffer capacity of the water and soil involved. While in the atmosphere, SO_2 and NOx gases and their derivatives (including sulfates, nitrates and ozone) contribute to visibility degradation and harm public health [40].

Once MNT-enabled solar energy becomes the exclusive energy source, problems such as acid rain and smog should not exist. In addition, future vehicles that are constructed from nano-materials, driven by nano-electro-mechanical systems and powered by hydrogen fuel cells (see Fig. 7) or solar cells (see Fig. 8) should totally eliminate transportation-related SO_2 and NOx emission. Therefore, the anthropogenic release of SO_2 and NOx that has assaulted the atmosphere since the Industrial Revolution should be ceased; further acidification of the environment and the threat to human health will be relieved [31–33].

Nanobots (such as nano-desulfurizer) could be sent up into the atmosphere to capture SO_2 gas, reduce it to sulfur and precipitate it back to the surface as dust. Nano-sulfur precipitators containing calcium or magnesium ions could be sent up into the atmosphere to oxidize SO_2 and then form $CaSO_4$ or $MgSO_4$ salts. Nanobots, like nano-catalytic converters, could also be sent up into the atmosphere to convert NOx into nitrogen and oxygen [31–33]. If the agricultural industry still needs fertilizer, nanobots like nano-NOx-reducers may be sent up to capture NOx and transform it into ammonia and bring it down to the ground. For ground level treatment of acidified water bodies and soil, a group of nano-buffers could be disseminated in them in order to increase their buffer capacity in resisting acidity. In addition, an army of nano-neutralizers may also be deployed to dynamically adjust the pH of water or soil to their original condition, either by capturing H^+ from the environment or by giving off OH^-.

4.2.2.2
Global Warming

The reason for global warming is the excess greenhouse CO_2 gas that has been dumped into the atmosphere due to fuel-burning. About 300 billion tons of excess CO_2 has been added to the atmosphere [40, 43]. Climatologists project that climbing CO_2 levels, by trapping solar energy, will partially melt the polar caps, raising sea levels and flooding coasts by the middle of this century [48, 49]. Applying MNT to energy production should enable solar power to be generated at an affordable cost. This will eliminate fossil fuel power generation as well as CO_2 emission; and thereby relieve concerns about a flooding catastrophe.

MNT could also be used to develop nano-machines or nanobots such as nano-photosynthesizers, nano-chlorophylls, and nano-carbon-fixers [31–33]. Powered by cheap solar energy, these nanobots could be manipulated to extract all of the 300 billion tons of excess CO_2 from atmosphere, and then to transform them into valuable materials. The carbon extracted by nanobots could be used to synthesize functional and structural materials [114, 118, 119, 124]. They could also be extracted by other nanobots and further synthesized into sugar, starch, and cellulose to supplement the demand for food, and paper. This could relieve the pressure exerted on farmland and forest.

4.2.2.3
Ozone Layer Depletion

The chief threat to the ozone layer comes from the release of CFCs [38]. Of the 682 million kilograms of CFCs consumed globally during 1991, 32% was used for refrigerants, 28% for blowing agents, 20% for cleaning agents, and 18% for propellants. Each molecule of chlorine in CFC is capable of degrading over 100,000 molecules of ozone before it is removed from the stratosphere or becomes part of an inactive compound [41]. Facing the destructive power of these ozone-depleting substances (ODS), the Montreal Protocol, an international agreement, was drafted by world governments to protect the ozone layer [38, 39].

The relative potency of the different ODS depends on the stability of the reservoir compounds. Bromine reservoirs (for instance HBr and $BrONO_2$) are 10–100 times more effective than chlorine at destroying ozone [38]. The nonreactivity of ODS allows them to drift for years in the environment until they eventually reach the stratosphere. High in the stratosphere, intense UV solar radiation severs halogens off the ODS, and it is these unattached halogens that are able to catalytically convert ozone molecules into oxygen molecules [39]. Different ODS are removed from the stratosphere in different timescales, ranging from 50–200 years. Despite the fact that the growth-rate of ozone depletion potential (ODP) in the atmosphere is starting to drop, without MNT the impact of ODS on stratospheric ozone will continue.

Future MNT will use water as solvent sparingly, and can recycle it 100% [31]. Even at this pre-breakthrough stage, ODS refrigerants can be replaced at a higher cost. MNT will help lower that cost to negligible. Therefore, the growth rate of ODP in the ODS reservoir will become zero. For the ODS remaining in the reservoir, Drexler [32] proposed using sodium-containing balloon type nanobots. The nanobots, powered by nano-solar cells, collects CFCs and separates out the chlorine in the stratosphere. Combining this with sodium makes sodium chloride. When the sodium is gone, the balloon collapses and falls. Eventually, a grain of salt and a biodegradable speck fall to Earth. The stratospheric CFC is quickly removed.

Extrapolating from Drexler's idea, nanobots containing other metals (like calcium and magnesium) can also be devised to remove stratospheric CFC. Among ODS, halogens other than chlorine (like bromine) could be neutralized using the same tactics (deploying an army of airborne solar energy-powered metal-containing nanobots into the stratospheric ODS reservoir). The originally expected 200 years' drag of ODS impact may be shortened to just a few years after the actual deployment date of the halogen neutralizing nanobots. Figure 11 is a conceptual idea of how the ODS reservoir in the stratosphere could be reduced by solar powered airborne metal nanobots, to relieve the ozone depletion problem.

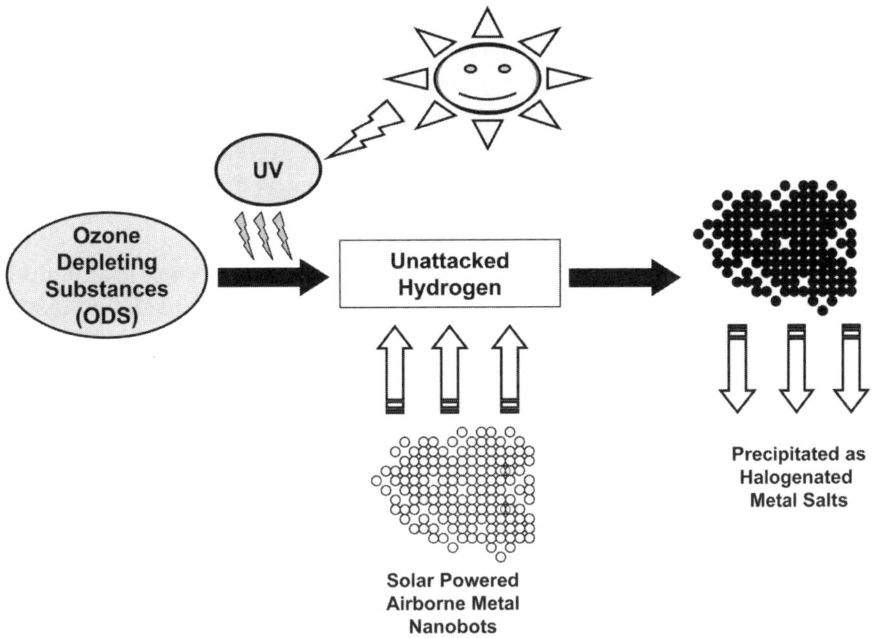

Fig. 11 MNT solution to stratospheric ozone depletion

4.2.2.4
Toxic Wastes

Toxic wastes, collected from different multimedia environments, can harm living systems. The contaminants in these wastes can be organic or inorganic [1, 2]. Organic contaminants (like PCB, Dioxin, see the chapter on "Forensic Investigation of Leachates from Recycled Solid Wastes") actually consist of harmless atoms arranged into noxious molecules. The inorganic contaminants contain toxic elements, such as lead, mercury, arsenic, and cadmium [16, 17, 41].

Contaminated dumpsites could easily be cleaned up using MNT. After identifying the chemical properties of the organic contaminants, specific nanobots for attacking specific contaminants at their weak points would be devised. An army of nanobots would then be deployed into the contaminated site to render the contaminants harmless by rearranging their atoms. The nanobots would be designed to be decayable, so no retrieval would be needed after the mission is completed [31–33]. They could also be designed to be reprogrammable, so that after each task they could be retrieved and reprogrammed for other cleaning tasks.

Inorganic contaminants may also be collected by specifically designed nanobots. Trillions of them could be sent into contaminated water bodies or soil to detect these toxic elements and store them inside. The nanobots would then be retrieved from the site, instructed to release the toxicants, and would then be ready for other tasks. The released concentrated metal ions could be rebuilt into rocks in the mines by nanobots or fused into a stable form by nanomachines and returned to nature [31–33].

5 Conclusions

In an industrial process, the transformation of raw materials into useful products is accompanied by the generation of waste. Apart from creating potential hazards and environmental problems, waste also represents a loss of valuable materials and energy from the production units. Billions of dollars are spent annually by industrial facilities worldwide to meet the discharge standards imposed by the local authority. Traditionally, pollution prevention has been achieved through treatment processes added at the end of the production line ("end-of-pipe"). However, this treatment approach is no longer viewed as adequate, because it does not actually eliminate wastes but simply transfers it from one medium to another. Increasing public awareness of the impact of industrial pollution, more stringent discharge standards, and escalating waste treatment and disposal costs have put enormous pressures on most industries to shift their pollution prevention from "end-of-pipe" treatment to waste minimization or even total elimination at the point of generation.

Recently, the U.S. Environmental Protection Agency (EPA) has declared waste minimization to be the top-priority action in the waste management hierarchy. Waste minimization primarily involves two main principles: source reduction and recycling. Source reduction includes changing the product through substitution or composition changes and controlling the waste at the source through input material changes, technology changes, and good operating practices (onsite and off-site recycling activities include direct use or reuse of the waste in the same or another process, and waste reclamation for other useful purposes). When implemented, the benefits from the waste minimization activity are enormous, and include: (a) economic benefits through cost savings in waste treatment and disposal, reduced raw material, energy, and utility usage, and increased process productivity and reliability; (b) health and safety benefits, by reducing the risk associated with handling hazardous materials; (c) legal benefits, by reducing the risk of breaching environmental regulations and the resulting liabilities, and; (d) benefits from an improved public image.

Despite the many benefits, systematic waste minimization in industry is only occasionally undertaken because of several issues perceived to hinder the successful implementation of a waste minimization program. Although reducing waste quantities brings down waste compliance costs and increases production revenue, the initial cost of implementing a waste minimization program is a significant economic deterrent.

With current state-of-art technologies, humans are still victimized by environmental pollution and threatened by hazardous wastes. Hence, environmental sustainability based on technology-to-date is pessimistic. Worldwide demand for environmentally friendly chemical processes and products requires the development of novel and cost-effective approaches to pollution prevention and total sustainability. One of the most attractive concepts for pollution prevention toward total sustainability is molecular nanotechnology (MNT). To survive with quality, there is a need for a real sustainable environment, and MNT has great potential to restore the environment to its original pristine state and obtain sustainability. Therefore, emerging MNTs, in all industrial fronts, offer radical tools that should allow human society, for the first time, to gain the upper hand in the struggle toward total environmental sustainability and economic growth.

References

1. Kassim TA (1998) PhD Dissertation, Department of Civil, Construction and Environmental Engineering, College of Engineering, Oregon State University, Corvallis, Oregon
2. Kassim TA, Simoneit BRT, Williamson KJ (2005) Forensic investigation of leachates from recycled solid wastes: An environmental analysis approach. In: Kassim TA, Williamson KJ (eds) Environmental Impact Assessment of Recycled Wastes on Surface and Ground Waters; Volume 3: Concepts, Methodology and Chemical Analysis. Handbook of Environmental Chemistry, Water Pollution Series, Vol 5/Part F. Springer, Berlin Heidelberg New York: 321–400
3. Brown ET (1990) State of the world: A World Watch Institute report on process toward a sustainable society. Norton, New York, pp 210
4. CBO (1985) Hazardous waste management: Recent changes and policy alternatives. Congressional Budget Office, Washington, DC
5. CMA (1989) Waste minimization resource manual. Chemical Manufacturers Association, Washington, DC
6. DHS (1986) Alternative technology for recycling and treatment of hazardous wastes: 3rd biennial report. California Department of Health Services, CA
7. Dufour JT (1989) Hazardous waste management handbook: Step-by step guide to environmental compliance. California Chamber of Commerce, Sacramento, CA
8. Environmental Defense Fund (1986) Approaches to source reduction of hazardous waste: Practical guidance from existing policies and programs. California Institute of Public Affairs, Claremont, CA
9. EPA (1987) Waste minimization environmental quality with economic benefit. Washington, DC
10. Hollod GJ, McCartney RF (1988) J Air Pollut Control Assoc 8:3–11
11. Hunt GE, Schecter RN (1988) Minimization of hazardous-waste generation. In: Harry M, Freeman A (eds) Handbook of hazardous waste treatment. McGraw-Hill, New York
12. MacLean RW (1987) Financial analysis of waste minimization options. In: CMA Waste minimization Workshop Proceedings, New Orleans, LA
13. OTA (1986) Serious reduction of hazardous waste for pollution prevention and industrial efficiency. Congressional Office of Technology Assessment, Washington, DC
14. Wyman R (1989) Waste minimization liability issues in solvent waste reduction. Alternatives USEPA, Cincinnati, OH

15. Librizzi WJ (1990) Hazardous waste treatment processes: Including environmental audits and waste reduction. Task Force on Hazardous Waste Treatment, Water Pollution Control Federation, Philadelphia, PA
16. Freeman HM (1990) Hazardous waste minimization. McGraw-Hill, New York, pp 254
17. Korenaga T (1994) Hazardous waste control in research and education. Lewis, New York, pp 427
18. Law Environmental (1994) Phase II RCRA Facility Investigation Work Plan, Alabama, pp 242
19. Martin WF, Levine SP (1993) Protecting personnel at hazardous waste sites. Butterworth-Heinemann (Elsevier, Amsterdam), pp 427
20. US EPA (1993) RCRA inspection manual. Office of Solid Waste and Emergency Response, EPA Reports, US EPA, Washington, DC, pp 317
21. US EPA (1990) Guides to pollution prevention: Research and educational institutions. Risk Reduction Engineering Laboratory, Center for Environmental Research Information, EPA Reports, US EPA, Washington, DC, pp 153
22. US EPA (1994) Reengineering RCRA for recycling – draft: Definition of Solid Waste Task Force, Office of Solid Waste and Emergency Response, US EPA, Washington, DC, pp 428
23. US EPA (1994) Setting priorities for hazardous waste minimization. Office of Solid Waste and Emergency Response, US EPA, Washington, DC, pp 113
24. Allen DT, Rosselot KS (1997) Pollution prevention for chemical process. Wiley, New York, pp 345
25. Christ C (1999) Chem Eng Technol 22:642–651
26. Crittenden BD, Kolaczkowski ST (1995) Waste minimization: A practical guide. Institute of Chemical Engineers, Rugby, UK, pp 429
27. Dantus MM, High KA (1996) Ind Eng Chem Res 35:4566–4578
28. Young DM, Cabezas H (1999) Comput Chem Eng 23:1477–1491
29. Yang Y, Shi L (2000) Comput Chem Eng 24:1409–1419
30. Treweek J (1999) Ecological impact assessment. Blackwell Science, London, pp 412
31. Drexler KE (1990) Nanotechnology. In: Yearbook of science and the future. Encyclopedia Britannica, London, 162:163–67
32. Drexler KE (1992) Nanosystems: Molecular machinery, manufacturing, and computation. In: Yearbook of science and the future. Encyclopedia Britannica, London, 170: 213–224
33. Drexler KE, Peterson C, Pergamit G (1991) Unbounding the future: The nanotechnology revolution, restoring the environment. William Morrow and Company, New York
34. Mulhall D (2002) Statement on nanoecology. Foresight Institute, Palo Alto, CA
35. Anastas PT, Williamson TC (1998) Frontiers in green chemistry. In: Anastas PT, Williamson TC (eds) Green chemistry frontiers in benign chemical syntheses and processes. Oxford University Press, Oxford, UK, pp 295
36. US EPA (1988) Waste minimization opportunity assessment manual. US Environmental Protection Agency Hazardous Waste Engineering Research Laboratory, Cincinnati, OH
37. Williams PT (1998) Waste treatment and disposal. Wiley, Chichester, UK, pp 419
38. Anderson JG, Brune WH, Proffitt MH (1989) J Geophys Res 94:465–479
39. Farman JC, Gardiner BG, Shanklin JD (1985) Nature 315:207–10
40. Graedel TE, Crutzen PG (1993) Atmospheric change: an Earth System perspective, 2nd edn. Freeman, New York
41. McFarland M, Kaye J (1992) Photochem Photobiol 55:911–29
42. National Science and Technology Council (1996) Committee on Environment and Natural Resources, national acid precipitation assessment program biennial report

to Congress: An integrated assessment. National Science and Technology Council, Washington, DC
43. Hilaly AK, Sikdar SK (1994) J Air Waste Manage Assoc 44:1303–1308
44. Isalski WH (1995) Environ Protect Bull 34:16–23
45. Shemwell BE, Levendis YA (2000) J Air Waste Manage Assoc 50:94–102
46. Field F, Kirchain R, Clark J (2000) J Ind Ecol 4:71–91
47. MacLean HL, Lave LB (2000) Environ Sci Technol 34:225–231
48. Climate Change (2002) Intergovernmental panel on climate change 3rd assessment report. Cambridge University Press, Cambridge, UK, pp 205
49. Durlak SK, Biswas P, Shi J-C, Bernhard MJ (1998) Environ Sci Technol 32:2301–2307
50. Cabezas H, Harten PF, Green MR (2000) Chem Eng 107:107–109
51. Bleier EA (1992) Soc Petroleum Eng 24553
52. Caudle A, Bansal S (1993) Soc Petroleum Eng 26010
53. Christainsen C (1991) Soc Petroleum Eng 23359
54. Degouy D (1991) Soc Petroleum Eng 23063
55. Friedheim EA (1991) Soc Petroleum Eng 23062
56. Longwell A, Akers TD (1992) Soc Petroleum Eng 23916
57. Retz RH (1991) Soc Petroleum Eng 23064
58. Thurber NE (1990) Waste minimization in E&P operations. Proc 1st Int Symp on Oil and Gas Exploration and Production Waste Management Practices. New Orleans, LA, 10–13 September 1990, pp 52–59
59. Thurber NE (1991) Soc Petroleum Eng 23375
60. Walker SH (1990) Soc Petroleum Eng 19525
61. Wojtanowicz AK (1991) Soc Petroleum Eng 21954
62. Wojtanowicz AK (1991) Soc Petroleum Eng 22815
63. Mitchell SD (1990) Soc Petroleum Eng 20436
64. Ogassawara FW (1991) Soc Petroleum Eng 22555
65. Traonmilin ED (1992) Soc Petroleum Eng 23876
66. Wojtanowicz AK (1993) Soc Petroleum Eng 26013
67. Beihoffer LS (1992) Oil Gas J 17:29–36
68. Burke H, Veil L (1995) Soc Petroleum Eng 29737
69. Clark R, Almquist A (1991) Soc Petroleum Eng 22550
70. Enright DP (1992) Soc Petroleum Eng 21937
71. Oakley DJ (1992) Soc Petroleum Eng 20885
72. Stilwell CT (1990) Soc Petroleum Eng 20713
73. Barragan KS (1992) Soc Petroleum Eng 23625
74. Beihoffer LS (1992) Soc Petroleum Eng 24588
75. Bland D, Ron A (1992) Oil Gas J 17:63–82
76. Bland WE (1992) Soc Petroleum Eng 24591
77. Edward-Berry FK (1992) Soc Petroleum Eng 24589
78. Friedheim EA (1991) Soc Petroleum Eng 21941
79. Geehan T (1991) Soc Petroleum Eng 23374
80. Rautela MS (1991) Soc Petroleum Eng 23834
81. Stringfellow AD, Jacobs M (1992) Soc Petroleum Eng 23920
82. Bushnell-Watson LS (1991) Soc Petroleum Eng 22549
83. Candler JE (1993) Soc Petroleum Eng 25993
84. Dachary L, Vighetto H (1992) Oil Gas J 17:51–62
85. Dearing HL (1990) Soc Petroleum Eng 19957
86. Hall RA (1992) Soc Petroleum Eng 24610
87. Halliday LS (1990) Soc Petroleum Eng 20461

88. Henriquez LR (1990) The development of an OBM cutting cleaner in the Netherlands. Proc 1st Int Symp on Oil and Gas Exploration and Production Waste Management Practices. New Orleans, LA, 10–13 September 1990, pp 21–23
89. Jacques WL (1992) Soc Petroleum Eng 23881
90. Harder GLL (1992) Soc Petroleum Eng 23928
91. Malachosky LK (1991) Soc Petroleum Eng 19528
92. Peresich EA (1991) Soc Petroleum Eng 21935
93. Schmidt E (1990) PRS treatment and reuse of oilfield wastewaters. Proc 1st Int Symp on Oil and Gas Exploration and Production Waste Management Practices, New Orleans, LA, 10–13 September 1990, pp 93–98
94. Zakharov M, Konovalov AA (1992) Oil Gas J 17:83–91
95. Foxenberg LW (1992) Soc Petroleum Eng 24784
96. MacEachern D, Young S (1992) Oil Gas J 17:132–145
97. Plank JP (1992) Oil Gas J 17:22–28
98. Reid PI (1993) Soc Petroleum Eng 25989
99. Sunde EL (1991) Soc Petroleum Eng 23343
100. Welch L (1992) Oil Gas J 17:101–112
101. Makoto A (2001) A future society built by nanotechnology. Science and Technology Group; Environment, Science and Technology Bureau, the Japan Federation of Economic Organizations (Keidanren), Tokyo
102. Wiesner M, Barron A, Colvin V (2002) Nano-structured membranes and their applications. Center for Biological and Environmental Nanotechnology, Rice University, Houston, TX
103. Das S (2000) J Adv Mater 52:41–44
104. Das S, Curlee TR, Jones DW, Leiby PE, Rubin JD, Schexnayder SM, Vogt DP, Wolfe AK (1999) Supporting infrastructure and acceptability issues for materials in new generation vehicles (ORNL/TM-13731). Oak Ridge National Laboratory, Oak Ridge, TN
105. Graedel TE, Allenby BR (1998) Industrial ecology and the automobile. Prentice Hall, Upper Saddle River, NJ
106. Hendrickson CT, Horvath A, Joshi S, Lave LB (1998) Environ Sci Technol 32:184A–191A
107. Kelkar A, Roth R, Clark J (2001) J Adv Mater 53:28–32
108. Lave LB, Cobas-Flores F, Hendrickson CT, McMichael FC (1995) Environ Sci Technol 29:420A–426A
109. National Research Council (NRC) (2002) Effectiveness and impact of corporate average fuel economy (CAFE) standards. National Academy, Washington, DC
110. Bruchez M, Moronne M, Gin P, Weiss S, Alivisatos AP (1998) Science 281:2013
111. Brune DA, Bicerano J (2002) Polymer 43:369–387
112. Garces JM, Moll DJ, Bicerano J, Fibiger R, McLeod DG (2000) Adv Mater 12:1835–1839
113. Basiuk EV, Rybak-Akimova EV, Basiuk VA, Acosta-Najarro D, Saniger JM (2002) Nano Lett 2:1249
114. Dwyer C, Guthold M, Falvo M, Washburn M, Superfine R, Erie D (2002) Nanotechnology 13:601
115. Gao H, Kong Y, Cui D, Ozkan CS (2003) Nano Lett 3:471–473
116. Mukherjee P (2001) Nano Lett 1:515
117. Taton T, Mirkin C, Letsinger R (2000) Science 289:1757
118. Cui Y, Wei Q, Park H, Lieber CM (2001) Science 293:1289
119. Dickson DPE (1999) J Magn Magn Mater 203:46
120. Lee K-B, Park S-J, Mirkin CA, Smith JC, Mrksich M (2002) Science 295:1702–1705
121. Ma J, Qi Z, Hu Y (2001) J Appl Polym Sci 82:3611–3617
122. Gardea-Torresdey JL (1999) J Nanopart Res 1:397

123. Gardea-Torresdey JL, Parsons JG, Gomez E, Peralta-Videa J, Troiani HE, Santiago P, Yacaman MJ (2002) Nano Lett 2:397–401
124. Gogotsi Y, Libera JA, Guvenc-Yazicioglu A, Megaridis CM (2001) Appl Phys Lett 79:1021
125. Hummer G, Rasalah JC, Noworyta JP (2001) Nature 414:188
126. Naiwa HS (2000) Handbook of nanostructural materials and nanotechnology. Academic, New York, vol 1–5
127. Office of Technology Assessment (1988) Advanced materials by design. U.S. Government Printing Office, Washington, DC, OTA-E-351
128. Grau R, Graells M, Corominas J, Espuna A, Puigjaner L (1996) Comput Chem Eng 20:853–863
129. Mulholland KL, Dyer J (1999) Pollution prevention: Methodology, technologies and practices. American Institute of Chemical Engineers, New York, pp 523
130. Alivisatos EA (2002) Science 29:2425–2427
131. Shapiro LE (2000) Phys Rev B 62:63–78
132. Stevens GB, Reda T, Raguse B, Braach-Maksvytis V, Tilbrook D (2000) Electrochemical reduction of CO_2 at a 200 nm porous Au film electrode in aqueous $KHCO_3$ electrolyte under near-ambient conditions. CSIRO, Australia
133. Sayama K, Mukasa K, Abe R, Abe Y, Arakawa H (2002) Photo-catalytic water splitting system into H_2 and O_2 under visible light irradiation mimicking a Z-scheme mechanism in photosynthesis. Solar Light Energy Conversion Team, Photoreaction Control Research Center, Advanced Industrial Science and Technology (AIST), Ibaraki, Japan
134. Hews-Taylor K (2002) Artificial photosynthesis, basic research in biomimetic applications: multilayer organic photovoltaic cell. CSIRO, Australia
135. Nijs JF, Szlufcik J, Poortmans J, Mertens RP (2001) Mod Phys Lett B 15:571–578

Subject Index

- page numbers set in normal type: Vol.1;
- page numbers set in *italics*: Vol.2;
- page numbers set in **bold**: Vol. 3

ACBFS 92
Accounting theory 7
ACGIH *138*
Acid rain **195, 220**
Action 17, 40, 45
–, absence/presence 18
Activated sludge process 221
Activity coefficient, liquid phase **5**
– –, octanol **8**
Acute toxicity test 162
ACZA 155, 367
Additives **164–167, 186**
Additivity index *251*
ADI *4, 17, 24, 25, 38, 148;* **110**
Administrative procedures 10
Admixture **178**
Adsorption **133**
Advection 247; *5, 11, 13;* **34, 36**
Advection-dispersion equation **70**
Aeration 71
Aggregate *45–48*
Aggregate manufacturing, lightweight *89*
Air *150*
Air, exposure 300
Air pollutants *139*
Algorithm 47
Alkylphenol polyethoxylates 225
Alternating direction implicit (ADI) *4, 17, 24, 25, 38, 148;* **110**
Aluminosilicates *76*
Aluminum *51–55, 58, 64, 76, 77, 81, 85, 86*
Ammonium acetate *72*
Ampholytes 337
Anhydride *103*
Anothite *97, 103*
ANOVA 356
Antibiotics *2, 3, 26, 27, 33, 34, 37*
APnEOs 225
Aquifer *63, 68, 74–78, 84, 91*
Aridisol 156; *48, 261*
Arsenic *72, 73, 77, 80, 85*
Artificial neural networks 365
Ash *63*
Asphalt cement 107
Asphalt concrete 153
Asphalt pavement 61, 82, 154
– –, reclaimed 82, 106, 128, 145
Asphaltenes 107
Assembler **209**
Assessment, biological *113*
Assessment factor *4*
Assessors 2
ATSDR *138*
Automation 344
Average daily intake *4*
AWQC *167*

Back-surfacer sand 85
Backward beam equation (BBE) **88**
Baghouse fines 67, 83, 90, 136
Bank protection 193
Barium *78*
Batch leaching 158, 279; *49*
Battelle environmental evaluation system 34, 38
BBDR *147*
Bearing capacity 304
Benchmark dose (BMD) *149, 167*
Benchmark response (BMR) *149*
Bentonite *90*
Beryllium *72, 73*
Bioaccumulation *74, 88, 167*
Bioassay *144*
Bioavailability **133**
Bioavailability modeling *213*
Biochemical oxygen demand (BOD) 220
Biodegradation 160, 283; *44, 45, 50, 57, 59;* **34, 134**

Bioleaching *71*
Biomarkers 325
Bioremediation 71
Biosolids 217
–, disposal, trends 228
–, surface applications 229, 230
Bituminous coal fly ash 94
Blast furnace slag 67, 90, 136
Boiler ash 79
Boiler slag 68, 93, 117, 137
Borderline substance *7*
Boring process *12, 17*
Boron *66–71, 91, 99*
Bottom ash 79, 154
Bound layers, fatigue 306, 314
– –, infiltration capacity 308–309, 314
– –, rutting 306
Breakwaters 193
Brick *62, 63, 92, 93*
Buffering 299
Bulk density **11, 12**
Bulk transport *205*

C&R 338, 366; *44, 240;* **68**
Cadmium *73–86, 99*
Caesium *76, 77, 83*
Calcite *75, 78, 83*
Calcium *64–67, 71–92, 99, 103*
Cancer slope factors (CSF) *145, 146, 155, 156, 166, 169*
Carbon *65, 67, 91, 92*
Carbon black **2, 11**
Carbonation 300, 315
Carcinogen *140–146, 163, 169, 181*
–, non- *142, 145, 157, 163–168*
Carcinogenicity *140, 141, 156, 168*
Carpet fiber dusts 67, 93, 117
$CaSO_3/CaSO_4$ 73, 95
Cathode ray tubes *173*
CB 209 332
Cellulose fibers 84
Cement kiln dust 74, 98
Cenosphere *92*
Ceramic *62, 92–96, 103, 104*
–, granules 85
CERCLA *177;* **67**
CFCs **220, 222**
Change 4
Channel flow 304, 310
Chemical fossils 325
Chemical oxygen demand (COD) 220
Chemical potential **3, 4, 13**
Chemodynamics *191*
Chemometrics 355
Chernobyl *88*
Chlorinated organics **134**
Chlorofluorocarbons **220**
Chlorophenols 333
Chromate *76*
Chromium *69–85, 99*
Chronic toxicity test 161
Circular economy 64
Clay *76, 80–86, 92, 93, 96, 103*
Clean Air Act *137, 167*
Cleanout slag 89
Clean-up techniques 343, 345
Climate change **196**
Cluster analysis 356
CO_2 **221**
Coal *64–68, 92–97, 103*
Coal bottom ash 68, 93, 117, 137
Coal fly ash 70, 94, 119, 137
Coal refuse 77, 100
Coal tars 70
Colloidal gas aphron **136**
Column leaching 158, 280; *206*
Cometabolism **43–46**
Composting 222, 223
Comprehensive strength **179, 180**
Computer recyclable materials *170*
COMs 367, *193*
Concrete *62, 63, 66, 68, 89, 93*
–, bituminous 84
Concrete aggregates 61
Concrete material, reclaimed 82, 107, 130, 145, 154
Concrete pavement, recycled *170*
Conductivity, hydraulic 241, 242, 252, 253, 256, 260–264
Construction and repair materials *112, 113*
Contaminant sourse location **65**
Contaminant transport **65, 69**
Contaminants **161–170, 175**
–, organic 326
–, –, toxicity *191*
Copper *72–84, 91*
–, slag 81, 104
Cordierite *96*
Cotransport **98**
Cracks 241, 242, 249, 253, 260–264, 300, 306–309, 314
–, roadway water movement 241

Cristobalite *97, 103*
CRMs **1**
Cullet 112

DAISY *10*
Damage 4
Daphnia magna 161, 276–278, 285; *44, 117, 119, 129*
Darcy's law **36**
DBT 334
1,3-DCB **146**
1,4-DCB **141**
Decision-making, environmental 11
Degradation, biotic *5, 6, 19–22, 32*
DEHP 332
Dehydrohalogenation **41–42**
De-icing salta 306–308
Dense nonaqueous phase liquids **100**
Denver potable reuse project *174*
Design facctor 11
Desorption *5, 11, 192, 261, 267;* **22, 23, 133**
–, isotherm **144**
–, kinetics **133**
–, resistance **150**
Dewatering, sludge 221, 222
1,4-Dichlorobenzene **140**
Dichloroethane **1, 2, 140**
Diffusion 247, 263; **36, 40, 176**
–, coefficient **14, 23, 25**
Digestion, aerobic/anaerobic 222
Dihaloelimination **47**
Discipline-orientedness 7
Discriminant analysis 356
Dispersion *5, 11;* **34–37**
Dispersivity, scale-effect **37**
DNAPLs **100**
Dose-response *139, 145, 147*
– –, biologically based *147*
DPT 334
Drainage pipes 194, 201, 206
Drainage system 305
Drilling fluids **204**
Drinking water *153–164, 169, 175, 180*
Driving force model, linear **27**
Dual-Monod kinetics **45, 46, 53–56**
Dump *63, 69–88, 104*
Dust, contaminated *151*

EAIA 3, 325
EBFS 92
Eco-development 5
Ecogenetics *139*
EDTA extraction *88*
Effect 4
EIA 1, 3, 207
–, leachates 209
–, characteristics 17
–, data 7
–, methods, desirable properties 21
Electron acceptors/donors **41–46, 52–55, 58**
Embryotoxicity *175*
Energy production **213, 215**
Energy recovery **201**
Environmental analysis and impact assessment (EAIA) 3
Environmental factors *114, 125*
Environmental forensic analysis 324
Epidemiologic studies *139*
Equilibrium distribution coefficient **10**
Equilibrium partitioning **1, 4**
Equilibrium relationship **9**
Equilibrium vapor pressure *213*
Erosion control 192, 197, 205
Explosives **47**
Exposure, acute *142*
–, chronic *143*
–, estimates, high-end (HEEE) *155*
–, pathways *151, 153*
–, point of *151*
–, reasonable maximum *154*
–, route *151*
–, subchronic *143*
Extraction 339
Eye irritation *142*

Factor analysis 358
Failure **49**
Fate-transport-toxicity model 152
Fatigue 306
Federal Water Pollution Control Act (CWA) *137*
FGD scrubber material 72, 95, 121
FGD sludge 73
Fick's Laws **14–18, 21, 24–27**
FID 371
Field contaminated soil **144**
FIFRA *137*
Fixation **160–164, 188**
Flame-retardant polymers *170*
Flat plate *45, 48, 49, 54, 55*

Flat plate leaching 158, 279; *244*
Flexural strength **177–180**
Flue gas desulfurization 72, 139
Fluoride *71*
Fluorogypsum 90, 110
Fly ash 79
– –, characterization *95*
– –, coal 61, 70, 94
– –, SEM *66*
Fossils, molecular/chemical 325
Foundry sand 74, 96, 122, 140, 154
Fractionation 344
Fractures 307
Freezing damage 311
Freundlich isotherm **9, 20, 39, 40**
FTIR 354
Fugacity **3–13, 23**
Fuller-Schettler-Giddings (FSG) equation **15, 16, 25**
Furnace slag 67, 88

Gahlenite *97, 103*
GANC *62, 94, 98–104*
Gas constant, ideal **3**
Gas exchange 309
GBFS 92
GC-MS 348
Genetic resource modeling 355, 387
Genotixicity *142, 146*
–, non- *142*
Germanium *83*
Gibbs free energy **3**
GIS *7*
Glass, postconsumer 89
–, waste, road construction 89, 112, 134, 149
Glass felt 84
Glass-formers 112
Goethite *77, 83*
Grain size *64*
Ground water 61; *150–153, 161, 162, 177;* **33, 65**
– –, intrusion 317
Grouting *43*
GUS *6, 7, 24–26, 32–36*

Haldane inhibition kinetics **43, 53, 54**
Half-life **41**
Halloysite *85*
Hazard index *141, 157*
Hazardous waste 63
Hazardous waste dumping **196**
Health Advisories (HA) *138, 165*
Hearth incinerator 85
Heat conduction **65**
Heavy metals 299, 305–308, 310; **160, 188, 196**
Hematite *67, 96*
Henry's law/constant *213;* **5–7**
Hereditary effects *142*
Hexachlorobutadiene (HCBD) **134, 144**
High surface area carbonaceous material **11**
Highway materials 61, 272, 282, 289, 322; *44, 111, 191*
HOCs **136**
HOPs **1**
Humic substances **113**
HWMs **1**
Hydraulic conductivity 241, 242, 252, 253, 256, 260–264
Hydraulic structures 184
Hydrocarbons 326
Hydrogenolysis **46, 47**
Hydrologic inversion **65**
Hydrolysis **41**
Hydrophobicity **8**
Hydroxides *77*

Identification 348
IDMS 339
Immobilization **160, 161, 166, 177**
Impact 4, 17, 40, 45
Impact assessment *63*
– –, data banks 9
– –, environmental 1
Impact estimation/indicators 19, 45
Incineration 223
Induction period *139*
Infiltration, roadway water movement 248, 251, 261–264
Infiltration capacity 308
Inhalation *142, 153*
Inhibition, competitive **44, 45**
Inspection procedures 26
Integrated Risk Information System (IRIS) *150, 169*
Ion-exchange *76, 79–81, 85*
Iron *64, 71–87, 90*
Isotherm, linear **10, 11, 20**

Joint action, independent *141, 158*
Joint toxic effect *230, 237*
Joints, roadway water movement 241, 242, 249, 261, 264

Kaolinite *76, 85*
Kiln dust 74, 98, 122, 140
Kinetics, first-order **41, 43, 46, 48, 56, 58**
–, Haldane inhibition **43, 53, 54**
–, modeling techniques *207*
–, Monod **53, 53, 56**
–, processes *205*

L/S variable 301
Land farming 71
Landfill *12, 17, 35, 62–70, 88–90*
–, gas 65
Landfilling 223
Langmuir isotherm **10, 11, 39, 40**
Lap granules 85
Laplace's equation **49**
LASs 225
LC-MS 354
Leachability **161, 164, 166, 168**
Leachates 156, 184, 338
–, chemodynamics *191*
–, forensic investigations 321
–, highway materials *127*
–, preparation 367
–, recycled SWMs *239*
–, soil *115*
–, waste and by-product materials *112, 113*
Leacher *7*
Leaching 157, 241, 245–253, 261–264, 293; *6, 7, 24, 25, 32, 35, 62–73, 93–98*
–, flux *240*
–, tests 295, 368
Lead *72–81, 84–90, 99*
–, slag 81
Leopold matrix 24, 26, 38
Lignite fly ash 94; *64, 93*
Lime kiln dust 74, 98
Liners, landfill *90*
Liquid-liquid extraction 341
Liquid-to-solid ratio (L/S) 247
LLE 341
LOAEL *148*
Local equilibrium assumption (LEA) **38**
LOEL *148*
Logit model *146*
Log-probit model *146*
Louisiana Superfund site **133**
Low water stream crossings 194

Mackinavite *78*
MACRO *2, 5–7, 11, 23–25, 36*
Macropore *11, 13*
Magnesium *64, 72, 81, 85, 90, 91*
Magnetite *67*
Maltenes 107
Manganese *71–87*
Manure *3, 12, 26, 34–38*
Marching-jury backward beam equation (MJBBE) **88**
Markers, biological 325
–, molecular 324, 358, 367, 373, 378
Mass transfer, interphase **1, 12**
Mass transfer correlations **98**
Matrix **160–169, 177, 181–187**
–, resolution 298
Maximum tolerance dose (MTD) *143*
MBT 334
MCIs *225*
Mercury *73–80, 86*
Metals, recyclable *170*
Micropore **11**
Microtox bacteria 162
MIKE 11 *10*
MIKE SHE *2–10, 15, 33, 36*
Mill tailings 77, 99, 154
Minamata disease *74*
Mineral processing wastes 76, 98, 123, 141
Minerals **2, 9**
Minimum relative entropy (MRE) **87**
Mission-orientedness 7
Mixture toxicity index *251*
Mixtures, unresolved complex 327, 373
Mobilization *63, 74–88*
Model process 46
MODFLOW **137, 154**
Modifying factors *148, 149*
MOLCONN-QSAR *251, 253*
Molecular connectivity indices *225*
Molecular fossils 325
Molecular markers 324, 358, 378
Molecular nanotechnology **191**
Molecular topography *225*
Mollisol 156; *48, 261*
Molybdenum *66, 74, 77, 91*
Monitored natural attenuation (MNA) **133**

Monod kinetics **53, 53, 56**
Monolithic block **167**
Monte Carlo 47
Montmorillonite *76, 81*
Mortar *68, 94*
MPT 334
MS, ionization techniques 349
MSW combustion ash 76, 98, 124, 142, 154
– – –, elemental composition 102
Mullite *67, 92, 96, 97, 103*
Multicomponent equilibria models *200*
Multicomponent joint toxic effect models *237*
Multicontaminant *193*
Municipal solid waste (MSW) 76

Nanobots **216, 223**
NAPL **97, 154**
–, pool **97**
NAPL-water interface **101**
Navier Stokes **25, 26**
NCHRP project 366
Neoplasms *175, 176*
Neptunium *88*
Nerve cells, simulated 365
Neurons, simulated 365
NHANES III *154*
Nickel *72, 79, 80, 83, 91*
–, slag 81, 104
NIOSH *138*
NOAEL *148, 149*
NOEL *148*
Non-leacher *7*
Nutrient inactivation *120, 126, 127*

Occlusion *83*
OCDD 230
Octanol water partition coefficient **8**
Oil shale, spent 79
One-hit model *146*
Organ effects *142, 175, 176*
Organic carbon analysis 372
Organic contaminants 225, 227, 232
– –, protocol 370
Organic mixtures, complex 348, 367
Organic solvents *173*
Organophosphorus 333
Organotin 334
Overlays 31, 38
Oxidation 299, 315; **42–46**
–, anaerobic **45–46**
Ozone hole **195**
Ozone-depleting substances **222**

PAEs 332
PAHs 228, 326, 385; *191, 194*
–, chronic toxicity prediction *253*
–, toxicity *252*
PAHs-QSAR *251*
Paint solvent reuse **208**
Paints **205**
Paper *171*
Partition coefficient *213*
– –, soil-water **139**
Pavement 241
Pavement structure 300–306, 311–316
– –, moisture content 300
– –, stress/strain 302
PBDEs 227
PBFS 92
PCBs 225, 331
PCDDs 225
PCE **86, 113**
PEC *3, 18, 25, 27, 32–37*
Pentachlorophenol 333
Perchloroethylene (PCE) **86, 113**
– pool **113, 118**
Percolation 247
PEST **127**
Pesticides 329; *7, 11–17, 22–25, 35*
Petroleum 70
–, hydrocarbons 326
Pharmacogenetics *139*
Phenols 330, 333
Phosphogypsum 90, 110, 154
Phosphorus *77, 91*
– slag 81, 104
Photobacterium phosphoreum 162
Photolysis 159, 282; *44–45, 58, 59*
Phthalates 332
Pipe bedding *43*
Pipe dope **205**
Pit design **204**
Pit slag 89
Plasticizer *44–59*
Plastics *170*
–, road construction materials 61, 82, 105, 126
PNEC *3–5, 17–24, 32–36*
Point source **66**
– – pollution **67**

Polanyi-Dubinin isotherm **10, 11**
Policy analysis 45, 48
Pollution of affluence/poverty 5
Pollution source identification **70**
Polybrominated diphenyl ethers (PBDEs) 227
Polychlorinated biphenyls (PCBs) 225–229
Polyethylene 82
Pond fines 83
Pore diffusion model **19**
Pore gas 300, 312, 315
Portland cement 84, 97; **160, 165, 177–180, 186–189**
Portland cement concrete (PCC) 113, 153; *43–46, 57*
Potassium *76, 81, 85, 86, 91*
Pozzolanas **160, 166, 186**
Pozzolanic *64–67, 95*
Precipitation 296, 310, 312
Pre-extraction 339
Preservation 339
Principal components analysis 357
Probability density function **79**
PVC 332
Pyrite *78, 86*

Q-mode factor analysis 358
QSAR *6, 191, 193, 224*
– modeling *213, 231*
QSPR *191, 193, 224*
– modeling *213*
Quarry by-products 82, 105, 127, 141
Quartz *67, 77, 96, 97, 103*
Quasi-boundary-value-problem (QBVP) **73**
Quasi-reversibility (QR) **73**

Rain collection systems 196, 203, 207
Raney-wells *74, 85, 86*
Rate laws *205*
Rate of release **146**
Receptor *152, 160*
Reclamation, offsite **201**
Recovery 338
Recycling **198**
Redox **42, 53**
Reefs, artificial 195, 202, 205
Reference concentration *145*
Reference doses *145, 148, 149, 157, 158, 168*
Reference state 18
Refuse-derived fuel 76
Regression equation constants *248*
Release 245–248, 263, 264
Remediation *63, 88;* **133**
Repair materials *112, 113*
Residues, leaching 293, 295
Resource Conservation and Recovery Act 113; *137*
–, factor **16**
Retardation 302, 311
Retardation factor **18, 20**
Revetments 193
Reynolds number **26**
RfD *168*
Rig-wash hoses **207**
Risk assessment 273, 286, 287; *3, 4, 23–27, 34, 36, 111, 125, 126*
Risk quotient *3–5, 36*
Road base construction *89*
Road construction materials 59
Road hydrology 300
Roadway culverts 194
Roadway water movement 241
Robotics 344
Roof replacement 84
Roofing shingle scrap 84, 108, 130, 146
RRR *43, 121*
– process 281
– – test methods 152, 157
Rubber, ground/crumb 85, 87
Rubidium *76*
Rutting 306

SACC 155
Safe Drinking Water Act (SDWA) *137, 162*
Safety factor *148*
Scarp tires, road construction 61, 85, 108, 131, 146
Schmidt number **26**
Schroeder increments **15**
Scrap tires 185
– –, composition 190
Screenings 83
Sediment quality *85*
Selenastrum capricornutum 155, 161, 276–278, 285; *44, 117, 119, 129, 251, 257*
Selenium *66, 73, 91*
Sensitivity analysis 47
Sewage sludge 217

Sewage sludge ash, chemical composition 109
– – –, road construction 85, 109, 133, 147
Sewage sludge treatment *91*
SFE 340
Sherwood number **26–28**
Shingle scrap 84, 108, 154
Shoulders, infiltration 309, 314
Shoulders, roadway water movement 241, 304
SIA 3
Siderite *78*
Silicate *62, 67, 76–86, 90–92*
Silicon *64, 81, 85*
Silver *73*
Simulation 40
–, validation 47
Single component system *194*
– – –, isotherm models *195*
Sintering *62, 92–98, 103, 104*
Skin irritation *142*
Slag, blast furnace 67, 90
–, copper/nickel/lead/zinc 81, 104
–, nonferrous 80, 103, 105, 125, 144
–, phosphorus 81, 104
–, steel, road construction 61, 88, 110, 134
Sludge 85; *3, 12, 35, 62, 91;* **161–169, 172, 177**
–, ash 109
–, dewatering/thickening 221, 222
–, EPA standards 224
–, fate 217
Smithsonite *83*
Smog **220**
SO_2 72
Social impact assessment (SIA) 3
Sodium *67, 85*
Soils *150, 151, 161, 261*
–, contaminated 70, 95, 121, 138
–, field contaminated **144**
–, metal-impacted 71
–, polluted *4*
Soil amendments 217, 223
Soil quality *85*
Soil sorption 159, 281; *117, 121, 127*
Soil stabilization *43*
Solid waste 63; **194**
Solid waste materials (SWMs) 62
– – –, highway applications 116
Solidification **159**
Solid-phase extraction 343
Solubility 247, 263; *213*
Solvents **205**
Sorption *267;* **2, 37–40**
Source confirmation 322
– –, risk assessment 324
Source medium *151*
Source partitioning model 322
Source reduction **191**
Speciation *5*
Stabilization 221, 222
Statistical theory 8
Steatite *96*
Steel slag, road construction 61, 88, 110, 134, 148
Stormwater sewers 195
Stream crossings, low water 194
Stream grade control 205
– – –, overfalls 192, 197
Stress/strain, pavement structure 302
Strontium *66, 78, 83–88, 99*
Structure-activity relationships *143*
Subbituminous coal fly ash 94
Subchronic studies *148, 166*
Sulfate wastes, road construction 89, 110, 134, 148
Sulfates *75*
Sulfides *75–84*
Sulfur *66*
Supercritical fluid extraction 340
Superfund Act (CERCLA) *177*
Surface waters 61; *150–153, 160, 161*
Surfactants 334
–, amphoteric 337
–, chemical structures 335
Sustainability **191**
SWMs *193*
–, leachates, molecular weights 374
–, recycled, organic leachates *239*
Synergistic action *141, 158*

Tap slag 88
TBT 334
1,1,2-TCA pool **111**
TCDD 230
TCE **86, 125**
Technetium *77, 88*
Tellurium *73*
Teratogenicity *175*
Tetrachloride **205**
1,1,2,2,-Tetrachloroethane **140**
Theoretical upper-bound estimates *155*

Thinners **205**
Thorium *88*
Threshold technique 33
Throughput economy 64
Tikhonov regularization **75**
Tiles *62, 63, 92*
Time-release history **65**
Tire block, bank protection 193, 199
Tire mattresses, lake/sea shore/harbor protection 200, 205
Tires, scrap, road construction 85, 108, 131, 146
Titanium *99, 103*
TOC *242, 245*
Top-down bulk technology **192, 210, 212**
Topology *225*
Tortuosity factor **16**
Toxic effect, joint *230, 237*
Toxic element release, cause *71*
Toxic Substances Control Act (TSCA) *137*
Toxic unit *251*
Toxicants *141*
Toxicity 160, 184
–, multicontaminant *191*
–, whole effluent *111, 113, 116*
Toxicity screening test 156, 161
Toxicity tests, acute/chronic 161, 162
Toxicology, genetic *139*
TPT 334
Tracers, organic 325
Transformation capacity **44, 45**
Transformation product toxicity **44, 45**
Transport, advective/dispersive **49, 54**
–, modeling fate **33**
Transport parameters, estimation *214*
Transport regimes 301
1,1,2-Trichloroethane **140**
Trichloroethylene **205**
Trihalomethanes **219**
Tumorigenicity *144*

Ultisol 156; *48, 261*
Unbound layers 304, 310–318
Uncertainty 30, 37, 42
Uncertainty factor *148, 168*
Unit risk factors *145, 146, 155, 156*
Unresolved complex mixtures 327, 373
Unsaturated flow 242
Uranium *73, 75, 77, 88*
USDA *154*

Value functions 25
Vanadium *69, 76*
Varimax 359
Vector rotation techniques 361
Volatile chemicals *151, 153*
Volatilization 159, 283

Wash slimes 78
Waste and byproduct materials 272, 273
Waste fibers 67
Waste glass 134, 149
Waste minimization **191**
Waste oils 70
Waste rock 77, 99
Wastes, organic chemical **196**
–, special 63
Wastewater treatment 85, 219
Water, exposure 295, 300, 315
–, intrusion 306, 309, 313, 317
–, lateral movement 301
Water quality 184; *63, 167*
Water-resource development projects 35
Weibull model *146*
Weight-of-evidence (WOE) *140*
Weighting, conventional 33
WET *111, 113, 116*
WHO *138, 148*
Whole effluent toxicity *111, 113, 116*
Wilke Chang equation **14**
WWTPs 219

XRD *67, 97*

Zeolite *72*
Zero-order reaction **48**
Zinc *72–86, 90, 91, 99*
Zinc slag 81, 104